U-BOATS AROUND IRELAND

THE STORY OF THE ROYAL NAVY'S COAST OF IRELAND COMMAND DURING THE FIRST WORLD WAR

GUY WARNER

COLOURPOINT

Published 2018 by Colourpoint Books
an imprint of Colourpoint Creative Ltd
Colourpoint House, Jubilee Business Park
21 Jubilee Road, Newtownards, BT23 4YH
Tel: 028 9182 6339
Fax: 028 9182 1900
E-mail: sales@colourpoint.co.uk
Web: www.colourpoint.co.uk

First Edition
First Impression

A catalogue record for this book is available from the British Library.

Designed by April Sky Design, Newtownards
Tel: 028 9182 7195
Web: www.aprilsky.co.uk

Printed by GPS Colour Graphics Ltd, Belfast

ISBN 978-1-78073-176-6

Front cover: Clockwise from top left – Royal Navy airship SS20, the US Navy destroyer USS *Nicholson*, German Navy U-boat U-10 and the Royal Navy battleship HMS *Audacious*.
Rear cover: Left to right – A flying boat receives attention at Queenstown, a depth charge from a US destroyer exploding, HM Trawler *Vera*.

About the author: Guy Warner is a retired schoolteacher and former civil servant, who grew up in Newtownabbey, attending Abbots Cross Primary School and Belfast High School before going to Leicester University and later Stranmillis College. He now lives in Greenisland, Co Antrim with his wife Lynda.
He is the author of more than 20 books and booklets on aviation and has written a large number of articles for magazines in the UK, Ireland and the USA. He also reviews books for several publications, gives talks to local history societies, etc and has appeared on TV and radio programmes, discussing aspects of aviation history.

Contents

Armed drifters in Larne Lough.
(Larne Museum & Arts Centre)

Foreword

Guy Warner's new book *U-boats Around Ireland* has evolved from his continued, extensive research into this subject, much of it at The National Archive at Kew, which has made it possible for him to significantly expand upon and widen the scope of the text in his earlier book entitled *Airships Over the North Channel* published in 2005. The result is a fascinating study of the Great War at sea from a perspective that has received insufficient attention from historians in the past. He describes the operations of small patrol vessels, minesweepers and lighter-than-air craft in the waters around Ireland that assumed critical importance during the Great War because they were used extensively by shipping heading to UK ports from the Atlantic and, thus, became the focal point of the enemy attack on shipping. This gave them strategic significance in addition to the local interest which first sparked Guy's enthusiasm and he has explained and put into context the composite nature of defensive operations by the Royal Navy, its Air Service and, later, the US Navy for the first time. In addition to its descriptions of operational matters, the book has a number of stories of human interest, not least the family connection with Guy's paternal grandfather who served at the RN air station at Larne and met his future wife there.

The material describing the arrival of US Navy ships and aircraft in the area from 1917 onwards is fascinating, as are the descriptions from German sources of what life was like for the enemy sailors in U-boats and from American sources about life in small destroyers. The appendices with their lists of warships that operated in Irish waters are a valuable reference source in their own right, as are the sobering statistics of merchant ship losses that accompany them. The intimate spirit of co-operation between the RN and USN in defeating the threat of U-boats and mines during 1918 was subsequently described by Franklyn D Roosevelt, Assistant Secretary of the US Navy at the time, as something that 'stands out as the finest example of the right spirit of co-operation between our two great countries'. The defeat of the enemy attack on shipping in Irish waters is an aspect of Great War naval operations that needs to be better understood and Guy has enabled a wider cross-section of the public to do so with this significant new book.

The extra text roughly doubles the size of the earlier book, fully justifying the change of title and I was pleased to see the emphasis placed in his description of how operations by patrol vessels and airships to achieve the common aim were integrated so successfully. The phrase that Admiral Jellicoe 'could have lost the war in an afternoon' in the North Sea is well known. From Guy's research we now know that had Admiral Bayly and the patrol forces under his command around the Irish coast not been as efficient and operationally successful as they were, defeat might have taken a few weeks longer but would have been just as certain. Thank you Guy, for producing such an informative and important book about events that took place a century ago and which deserve to be more widely remembered.

Commander David Hobbs, MBE
Royal Navy (Ret'd)
Crail, June 2018

Author's Foreword

IN 2005 THE FIRST edition of *Airships over the North Channel* was published. It examined the activities of the Royal Naval Air Service 'blimps' which, between 1915 and 1918 operated from a mooring out station that had been established at Bentra in Co Antrim, near the little seaside town of Whitehead. It was the first military aviation installation in Ireland and came under the command of the Senior Naval Officer at Larne Naval Base. My interest had been sparked by a family connection.

Eric Lionel Warner and Martha Knox both served at Larne Naval Base during the First World War. My grandfather was an Englishman from Worcestershire and was a Marconi-trained wireless telegraphist who served all round the world in the Merchant Navy in the years before the war. He became a Commissioned Warrant Officer in the Royal Naval Reserve.

After service in the armed merchant cruiser HMS *Macedonia*, including participation in the Battle of the Falkland Islands on 8th December 1914, he was posted to Larne as an instructor in wireless telegraphy. My grandmother was a country girl whose father was prominent in the Methodist Church at Ballynure. The family later moved to Station Road in Larne. During the war Martha joined the Women's Royal Naval Service – the Wrens, following the example of her older sister, Minnie, who was a Leading Wren. My grandparents met in the course of their duties and fell in love. After the war ended my grandfather was offered a permanent commission in the Royal Air Force. This he declined as he wished to stay in Larne and marry my grandmother – which they duly did on 25th October 1919 – who did not want to leave the town, remaining in the town for the rest of their long lives. They had two sons, my uncle Jim, born in 1921 and my father, Sam, who was born in 1924. This account is dedicated to their memory.

The first edition was well-received and in 2012, a second, expanded and revised version was published, re-titled *Airships over Ulster*. Over the six years between 2006 and 2012 I had given many talks on the subject of the airships to historical societies, military history

Below left: Some of HMS *Macedonia's* crew taking a little R&R – my grandfather is seated on the far left. *(Author's Collection)*

Below right: HMS *Macedonia* Xmas Party 1916. *(Author's Collection)*

Above left: Eric Lionel Warner and Martha Knox on their wedding day on 25th October 1919. *(Author's Collection)*

Above right: (L–R) Martha Knox WRNS, not known, Section Leader Minnie Knox WRNS. *(Author's Collection)*

societies, Probus and Rotary Clubs, church groups and others in Northern Ireland, the Irish Republic and Scotland. The universal reaction was one of great interest in this small piece of aviation, naval and local history. I was very gratified by this and also by the feedback from audiences, who increased my knowledge of the subject with personal memories. One elderly gentleman, Mr Billy Noble, at the age of almost 100, could remember the airships flying over Islandmagee – great, grey, silent shapes. Another lady, Vera Girvan, was able to tell me that her family home in the 1950s was one of the airmen's huts at Bentra and gave me a photograph to prove it.

By this means and by further research at the National Archives, Kew, I was able to add a little more to the text.

Over the last five years I have been able to delve into the National Archives again and certainly have proved (at least to myself!) the more times you do something the better you get at it. The result is that I have uncovered a considerable amount of additional information about the airships. I have also dug deeper into the details of the U-boats, their victims, the Royal Navy's response from Larne, Belfast, Londonderry, Queenstown, Kingstown, Berehaven, Buncrana, Rosslare, Killybegs and Rathmullan, to name just some of the principle bases, as well as the vital contribution made by the US Navy; setting the story in its wider overall context. The complex operation devised to combat and defeat the deadly U-boat menace by the RN, RNAS and in its later stages the RAF, USN and USNAS, certainly gives me the firm conviction that if the war had continued into 1919, the Allies would have defeated the Germans just as decisively at sea as in France. There would have been no 'stab in

Above left: Minnie Knox (at bow) at Larne Naval Base during WW1. *(Author's Collection)*

Above right: The huts used for accommodation and later as family homes at Bentra. Note the horseshoe above the door and the washing line. *(Vera Girvan)*

the back myth.' It is well known that Churchill described Admiral Sir John Jellicoe, C-in-C of the Royal Navy's Grand Fleet at the Battle of Jutland in 1916, as the only man on either side who could have lost the war in a day. The C-in-C Coast of Ireland, Admiral Sir Lewis Bayly, served at Queenstown for four years and could have lost the war if his command had not defeated the U-boats. If the sea approaches to Britain via the northern and southern coasts of Ireland had not been defended so capably and resolutely, the interruption or stoppage of the flow of vital materials (and from 1917, men) across the Atlantic would have resulted in defeat for the Allies. Indeed Admiral Eduard von Capelle, the Secretary of the Imperial German Navy from 1916 to 1918, regarded the northern and southern entrances to the Irish Sea and the western approach to the English Channel as, 'the decisive U-boat theatres, the highways of the world's traffic.' The text has, therefore, once more been expanded, this time to a considerable extent and many extra illustrations have been added. This now to my mind justifies a further change of title.

As ever, this work would not have been possible without the forbearance and support of my wife, Lynda. I am very grateful to my good friend, Ernie Cromie, for proof-reading the text and also to Commander David Hobbs for his support and encouragement.

Guy Warner
Carrickfergus 2018

Note on Ranks and Measurements

Ranks and Equivalents

Imperial German Navy	Royal Navy/US Navy
Grossadmiral	Admiral of the Fleet/Fleet Admiral
Admiral	Admiral
Vizeadmiral	Vice Admiral
Konteradmiral	Rear Admiral
Kommodore	Commodore
Kapitän-zur-See	Captain
Fregattenkapitän	Junior Captain (no direct equivalent)
Korvettenkapitän	Commander
Oberleutnant-zur-See	Lieutenant
Leutnant-zur-See	Sub Lieutenant/Lieutenant (jg)
Oberfänrich-zur-See	Midshipman/Ensign
Hauptbootsmann	Senior Chief Petty Officer
Oberbootsmann	Chief Petty Officer
Bootsmann	Petty Officer First Class
Obermaat	Petty Officer Second Class
Maat	Petty Officer/ Petty Officer Third Class
Hauptgefreiter	-
Obergefreiter	Leading Seaman/Seaman
Gefreiter	Able Seaman/Seaman Apprentice

Nautical Measurements

Distance: Cable: 1 cable = 183 metres (200 yards)

Distance: Nautical mile: The length of a nautical mile differs slightly in different latitudes but can be taken as 1853 metres (6080 feet) or, for practical purposes, 1829 metres (2000 yards).

Depth: Fathom: 1 fathom = 1.83 metres (6 feet)

Speed: Knot: Speed is expressed in knots: nautical miles per hour

Guns

Royal Navy Auxiliary Patrol Vessels

A 3-pounder gun fired a projectile weighing approximately 3.3 pounds (1.5 kilograms) with a calibre of 1.850 inches (47 millimetres) and a maximum firing range of 4000 yards (3657 metres).

A 6-pounder gun fired a projectile weighing approximately 6 pounds (2.7 kilograms) with a calibre of 2.244 inches (57 millimetres) and an effective firing range of 4000 yards (3567 metres).

A 12-pounder gun fired a projectile weighing approximately 12.5 pounds (5.66 kilograms) with a calibre of 3 inches (76 millimetres) and an effective firing range of 9300 yards (8503.92 metres).

A 7.5 inch (191 millimetres) naval howitzer fired a projectile weighing 100 pounds (45.4 kg) with a maximum firing range of 2100 yards (1920 m).

Imperial German Navy U-boats

A 50 millimetre (1.97 inch) deck gun fired a projectile weighing 3.86 pounds (1.75 kilograms) with an effective range of 5290 yards (4840 metres).

An 88 millimetre (3.5 inch) deck gun fired a projectile weighing 30.4 pounds (13.8 kilograms) with an effective range of 8000 yards (7300 metres).

A 105 millimetre (4.1 inch) deck gun fired a projectile weighing 51 pounds (23.3 kilograms) with an effective range of 16600 yards (15175 metres)

Map of locations
Malin Head
Ballyliffan
Lough Foyle
Rathlin Island
Tory Island
Lough Swilly
Buncrana
Ballycastle
North Channel
Ailsa Craig
Lough Foyle Naval Air Station
Rathmullan
Cushendall
Derry/Londonderry
The Maidens
Larne
Bentra/Whitehead
Belfast Lough
Killybegs
Aldergrove
Lough Neagh
BELFAST
Donaghadee
Strangford Lough
Portaferry
Skerries
Irish Sea
Malahide
DUBLIN
Dalkey Island
Dún Laoghaire
Galway
Limerick
Tralee Bay
Wexford
Johnstown Castle
Rosslare
Tuskar Rock
St George's Channel
Cork
Queenstown
Aghada
Berehaven
Castletownbere
Whiddy Island
Bantry Bay
miles
0 25 50 75 100
0 50 100 150
kilometres
Map data: ©MAPS IN MINUTES™/Collins Bartholomew 2007

Introduction

THE GREAT WAR OF 1914–18 brought many technological innovations which served to add to the horror and carnage of the conflict. The war at sea was not immune from these developments as battle was engaged not only on the surface but also by underwater craft and in the air above. Admiral of the Fleet Earl Jellicoe later wrote, 'German naval officers in pre-war days had not realised the influence that submarines would exert upon naval warfare, as the capabilities of the new arm to undertake oversea operations at long distances from their bases, and unescorted by surface vessels, had not been fully appreciated. Similarly the Royal Navy had looked upon the submarine as a vessel dangerous to surface craft, but possessed of but a limited range of action. The possibilities of the submarine as an offensive weapon came, therefore, as somewhat of a surprise to both sides.'

Admiral of the Fleet Lord Jellicoe. *(Library of Congress)*

A young officer, Leutnant zur See Johannes Spiess, who had, with some reluctance, been posted to the submarine service from the pre-Dreadnought battleship, SMS *Pommern*, had been hoping for assignment to a much more dashing torpedo boat and could not hide his dismay:

> 'In those days we looked at under-sea craft, along with aircraft and other technical innovations, with a sceptical eye. Would they ever amount to anything in real warfare? Probably not. Nor was life aboard the U-boats anything to look forward to. Even now the submarine is no pleasure barge. In 1912, between close quarters, foul air, and crazy rolling and pitching, a rowboat was palatial compared to the inside of one of those diving dories. There were frequent accidents, too, especially in foreign navies. And death in a plunging submarine was as evil a fate as the imagination could conjure. Death by slow suffocation. Nevertheless, although I did not like it, a submarine officer I became.'

Nowadays we can look back with an indulgent smile upon that prehistoric era. Any kind of extended U-boat voyage was undreamed of. Only in rare cases did men sleep on board, which was not only uncomfortable but considered dangerously unhealthful. Going ashore at nightfall was the invariable routine. Diving was done as little as possible, and we seldom ventured to go down more than a few yards, and then we looked anxiously about to see if the seams were tight and no water was leaking in. There was grave doubt whether sub-surface craft could weather a lively storm. They had never been tried out in a real gale. An attack under water in any kind of rough weather was considered impossible. The prescribed plan under such conditions was to approach and torpedo an enemy craft with the conning tower above water. The supposition was that, with the waves breaking over the conning tower, it could not be seen.' However, following a praiseworthy performance in naval manoeuvres in May, in December 1913, his boat, U-9, which was commanded by Kapitänleutnant Otto Weddigen, stayed out at sea in a violent storm 'and even carried out manoeuvres in the teeth of a gale. Decidedly the submarine as an instrument of war was picking up and beginning to give a hint of good possibilities.'

During the early months of the war, attacks on warships of the Royal Navy by submarines or U-boats of the Imperial German Navy concentrated on the Grand Fleet in the waters near its anchorage at Scapa Flow in the Orkney Islands or along the east coast of Scotland, where there was another naval base at Rosyth. However, as 1915 began, the area of hostile operations began to spread into the English Channel and the Irish Sea. In northern waters submarines were all but invisible when fully submerged, their presence given away only by oil patches on the surface. When at periscope depth and surveying the scene as they prowled below the waves, their position could then be betrayed by the distinctive but hard to spot (except from the air) 'feathering' of the periscope through the water.

In 1914 Ireland was a naval backwater with only one base of any size, at Queenstown in Co Cork – which was really a rest home for a senior, semi-retired admiral.

War with Germany was coming but it was anticipated that the naval engagements would

Above: A periscope viewed from above feathers through the water giving away its position. *(Airship Heritage Trust)*

Below: A sistership of U-9, U-7 at full speed, the bow of U-12 is also visible. *(Library of Congress)*

A pre-1914 image of Queenstown Harbour. *(Library of Congress)*

take place in the North Sea and around the globe in the scattered outposts where two empires clashed. However, by 1918 there were eighteen naval warship and naval air bases in Ireland, under the command of a full and very active admiral. Around the coast were ranged the forces of three navies, on the one hand the U-boats looking for prey and on the other battleships, cruisers, destroyers, sloops, minesweepers, patrol craft, submarines, depot ships, a host of requisitioned smaller vessels (trawlers and drifters), flying boats and airships. How and why did this come about?

Buncrana in the early 20th century. This little town, on the east shore of Lough Swilly, would become a naval headquarters. *(Author's Collection)*

Chapter 1
Larne, Kingstown and Lough Swilly 1914–15

Larne

In 1914 Larne was a thriving market town and seaport with a population of around 9000, 'a busy and important town situated at the entrance to Larne Lough.' Its situation on the north-east coast of County Antrim gave it a strategically important position on the North Channel between Ireland and Scotland, at the northern outlet of the Irish Sea where it meets the Atlantic Ocean. The harbour was described as, 'one of the safest and most commodious havens on this coast and always presents a scene of activity and animation, indicative of the extent and importance of the local trade.' Any ship using the Western Approaches had to travel either to the north or south of Ireland. At that time, of course, all of Ireland was part of the United Kingdom. The sea route from Stranraer, at the head of Loch Ryan, to Larne was the shortest crossing between Great Britain and Ireland.

Kingstown Harbour in 1899. *(Author's Collection)*

Kingstown

Kingstown, on the south side of Dublin Bay, had a population of about 18,000. It had been a fishing village until work on the harbour was begun in 1817. Its name was changed from Dunleary following the visit of King George IV in 1821. Ireland's first railway, from Dublin to Kingstown, opened in 1834. In the Victorian period it became the mail packet station between Dublin and Great Britain and the chief port for the steamers plying to Holyhead and Liverpool. It was also a popular seaside resort and yachting centre. It would be developed into a significant naval base.

Early naval developments in Larne

From the very beginning of the war in August 1914, Larne increased in importance as a port for embarking and disembarking troops and supplies, as well as becoming a naval base in its own right. The Larne to Stranraer ferry service had been operated by the Larne and Stranraer Steamboat Company since 1872. It became the Portpatrick and Wigtownshire Railways Joint Committee in 1890. The SS *Princess Maud* was a handsome, two funnelled vessel.

The handsome two-funnelled *Princess Maud. (Mid and East Antrim Borough Council)*

Her companions, the SS *Princess May* and SS *Princess Victoria*, were requisitioned by the Admiralty, leaving the *Maud* to maintain this important lifeline on her own for the duration. HMS *Princess Victoria*, which was a more modern ship, served as a seaplane tender and air service transport in the North Sea, while *Princess May*, a paddle steamer, had a more humble role as an accommodation ship. Due to the threat likely to be imposed by hostile submarines and floating mines, countermeasures of a sort were taken from the start. Sailings were made at irregular times and route deviations were introduced sporadically. As Larne developed as a naval base, it became possible to provide an escort. A steamer, loaded with troops and sailors, would have been a fine kill.

Lough Swilly

Lough Swilly is a quiet deep-water fjord in County Donegal and had been used by the Royal Navy as a fleet anchorage since Napoleonic times. Between 1798 and 1800 forts were built at opposite sides of the channel at Dunree, Ard's Point, Salpan Hill and Inch Island to the East and Macamish Point and Knockalla to the West. Knockalla and Dunree were the main forts, being closest to the mouth of the Lough and at its narrowest point. A typical event of the Victorian era was reported in the Belfast newspaper, the Northern Whig:

> 21st July 1899 CHANNEL SQUADRON IN LOUGH SWILLY Derry, Thursday
> This evening 27 warships, comprising the 'A' squadron of the Channel Fleet, unexpectedly entered Lough Swilly and came to anchor below Buncrana in three-line formation. Vice Admiral Sir H Rawson is in command, with Rear Admiral Fanshawe second. The vessels include the *Majestic*, *Magnificent*, *Jupiter*, *Repulse*, *Resolution*, *Prince George*, *Mars*, *Hannibal*, *Arethusa*, *Andromeda*, *Thames*, *Mersey*,

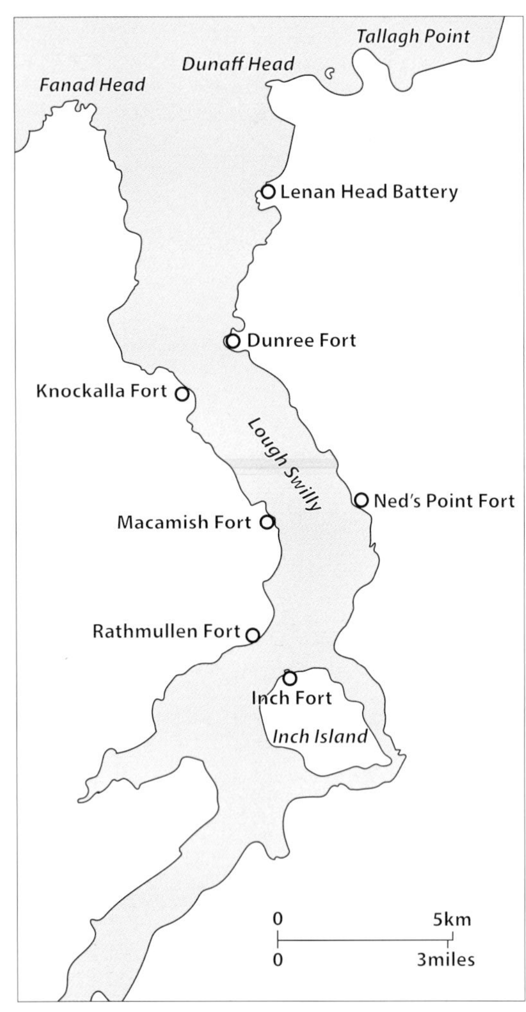

Above: The forts protecting Lough Swilly.

Below: The crew of HMS *Audacious* take to the lifeboats. *(Edith and Mabel Smith, licensed under the Creative Commons Share Alike 4.0)*

Diadem, *Furious*, *Niobe*, *Argonaut*, *Sybil*, *Pelorus*, *Minerva*, *Latona*, *Aeolus*, *Cambrian*, *Naiad*, *Fox*, and *Retribution*. Failing to enter Blacksod Bay on account of a fog, the squadron sailed north to Lough Swilly. Three vessels leave to-morrow, and the remainder on Saturday morning for Belfast Lough.

By the beginning of the First World War massive fortifications with gun emplacements had been built at both Lenan and Dunree.

In the autumn of 1914, while the anti-submarine defences at Scapa Flow in the Orkney Islands were being strengthened, many of the Royal Navy's Grand Fleet of dreadnought battleships were temporarily based at Lough Swilly between 22nd October and 1st November. Lough Swilly was selected as suitable by Admiral Jellicoe as it had a comparatively narrow entrance and shallow water, making it difficult for a submarine to enter undetected. He wrote, 'for the first time since the declaration of war, the fleet occupied a secure base'. People living along the lough shore witnessed the arrival of some warships. Fresh water was piped from the hills and supplies were bought in from local towns and villages. Admiral Jellicoe divided his time between Buncrana on the eastern shore and the seaside village of Rathmullen across the water. A boom was subsequently placed across the lough between Macamish Point and Ned's Point to further improve its security. However, while the Grand Fleet was there, the new battleship, HMS *Audacious*, under the command of Captain Cecil Dampier, was sunk on 27th October by a mine fortuitously

The new battleship HMS *Audacious* prior to striking a mine off Tory Island. *(Library of Congress)*

and quite coincidentally laid off Tory Island by the German auxiliary minelayer, SMS *Berlin*, commanded by Kapitän zur See Hans Pfundheller. The Flag Officer of the First Battle Squadron was Vice Admiral Lewis Bayly. He insisted on boarding *Audacious* to assess the situation for himself and was the second last to abandon ship, followed by the captain. Another ship, the SS *Manchester Commerce*, had been sunk by one of *Berlin's* mines on the day before, with the loss of 14 lives – the first ship to be sunk off the coast of Ireland in the war.

As a direct result of this catastrophe, mine-sweeping trawlers were sent to Lough Swilly and also to Larne. The little town of Buncrana, population 2000, became the naval headquarters. It was some 14 miles from Londonderry and was connected by the Londonderry and Lough Swilly Railway. The author of Black's Guide to Ireland waxed lyrical about its charms, 'No one who stays at Buncrana in good summer weather, when the scenery may be seen at its best, will regret it. The spot is charming. Built on a beautiful bend of the winding Swilly the 'Lough of Shadows' and engirdled with striking mountains, this bit of Donegal is a treasure of which the sons of Erin may with reason be proud.'

Of Londonderry, population 40,000, it remarked:

> '...it is one of the prosperous-looking towns of Ireland. It is bright and clean and its main thoroughfares wear a decidedly business-

A pre-1914 view of Londonderry. *(Library of Congress)*

An armed trawler. *(Author's Collection)*

like air. The town is an important seat of the linen manufacture, but the staple is shirt-making, which employs more than 20,000 hands, mostly female. It also possesses shipbuilding yards, iron foundries, distilleries, and breweries. The harbour is commodious, and a very large coasting trade is carried on.'

Larne

The Olderfleet Hotel in Larne became a naval headquarters under the direction of the Senior Royal Naval Officer, who in 1914 was Commander Hastings GF Berkeley. In the early months of the war the armed yacht *Oriana* and four drifters, *Boy Scout*, *Dick Whittington*, *Ocean Harvest* and *Ocean Retriever*, the 'Northern Patrol' under Lieutenant Commander AR Pack, were supplemented by six tug boats, hired from the Alexandra Towing Company of Liverpool, *Alexandra*, *Blackcock*, *Harrington*, *Herculaneum*, *Hornby* and *Wallasey*. These constituted the 'Southern Patrol', commanded by Lieutenant Commander AO Morgan.

HM Armed Drifter *Daisy VI*, seen here at Berehaven. *(Author's Collection)*

A similar number of tugs was based at Moville on the north western side of Lough Foyle, 'a clean, pleasant town, finely sheltered by high hills from the western gales and possessing a good beach for bathing, where the American mail steamers of the Anchor and Allan Lines stop once a week on their way between Glasgow and New York.' In Larne, the parent company in Liverpool was under contract to supply victuals, coal and 'other necessary stores', while for the Moville-based tugs this had been sub-contracted to MacDevitte and Donald of Foyle Street, Londonderry. While the tugs could patrol for 8–11 days at a time they were not very suitable for the inspection of suspect vessels as their large rubber fenders impeded the boarding parties. Communications were also a problem as the wireless telegraphy link was unreliable, requiring a tug to be dispatched with a message. Marconi Company engineers were sent over from England to see if they could make urgent improvements. A sizeable fleet of armed trawlers and drifters, sturdy vessels for patrolling the sea lanes, began to assemble at the Naval Base. The first of these welcome reinforcements to arrive were the trawlers, *Nellie Braddock*, *Davara*, *Fishtoft*, *Angerton*, *Alsatian*, *Riano*, *Cerealia*, *Revello* and *Liberia* in November and December 1914 to replace the tugs. Typically the larger trawlers were fitted with 6-pounder guns and the drifters with 3-pounders, as soon as these could be made

'Sturdy little ships, manned by men of the finest type.' A typical Armed Trawler. *(Author's Collection)*

available. The mine-sweeping trawlers *Kaphreda*, *Loch Doon*, *Athelstan* and *Yucca* were sent from Larne to Buncrana on 27th November to be based there under the command of Lieutenant Sir James Domville, Bart. The need for these was emphasised on 20th December when SS *Tritonia* was blown up by a mine 22 miles NNE of Tory Island.

Winston Churchill would later write, 'There is no more brilliant chapter in the history of the Great War than that outlining the story of the armed trawlers and similar vessels. These sturdy little ships, manned by men of the finest type, carried out the dangerous work of minesweeping, in addition to doing duty in patrol work and convoy escort. At sea at all times and in all weathers they accomplished their task in accordance with the highest tradition of the Royal Navy and the Merchantile Marine.' To which may be added former private pleasure yachts and excursion paddle steamers, similarly taken up and crewed by the RNR and RNVR. E Keble Chatterton, who served at both Queenstown and Berehaven, later becoming a prolific nautical historian, wrote, 'Trawler skippers, who had done nothing all their lives except fish, found themselves in positions of the highest responsibility; deckhands who had never so much as fired a rifle became experts with a 12-pounder mounted on the trawler's fo'c'sle. Junior officers out of some big liner found themselves in command of a whole division of trawlers, putting to sea in a single line ahead and then separating along the coast.' He added, 'You can lead a fisherman but if you try to drive him you only put his back up. As time went on naval officers began to realize what a magnificent, plucky, hard-working, resourceful body of men these were.' They did, however, need to be treated with much greater tact and latitude than contemporary naval discipline then allowed:

> 'I have no wish to depict the trawlermen of the new Navy as the embodiment of all virtues. This they certainly were not. The career of a fisherman sometimes consists of a strenuous time at sea, punctuated by a series of 'rough-house' incidents in harbour. The trawlerman of 1915 was a rough, tough, hard-case fellow who did

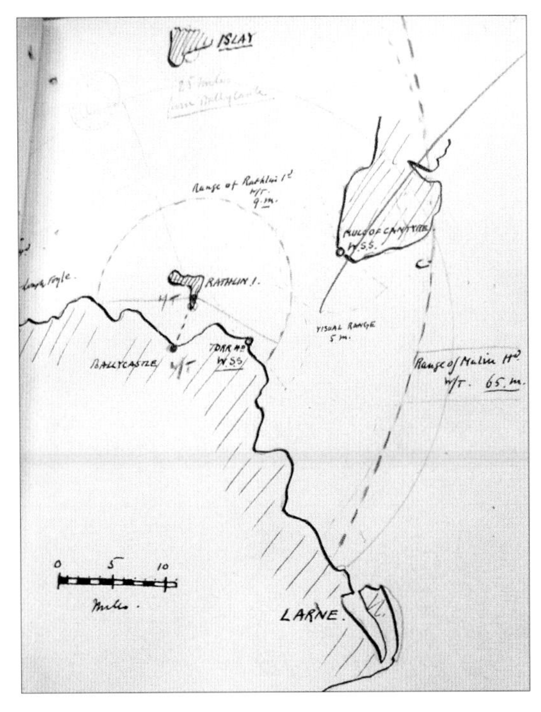

A sketch map drawn in 1915 illustrating the range of radio transmitters in the North Channel. *(National Archive)*

not take kindly to naval discipline at first. On the other hand many naval officers foolishly attempted to use the methods on independent fishermen which had been devised only for those accustomed to the life and discipline of the Royal Navy.'

The route around the north coast down towards the Clyde and the Mersey was a fertile hunting ground for the enemy and would become ever more so as the war persisted and became a desperate struggle for national survival. More than 4000 fishing vessels and 159 steam yachts were taken up by the Admiralty during the war.

While based at Larne, these small ships would have their coal bunkers replenished between patrols, their boilers would be cleaned, repairs made, the crews rested, supplies would be taken on board and the necessary command and control would be given. Further trawlers arrived and were placed in three divisions numbered Units 68, 69 and 70, ideally with six in each. It was recognised that the extra vessels would put a strain on facilities, resources, quay capacity and manpower. A much needed enhancement was effected on 16th January when the wireless telegraphy station at Ballycastle was brought on line.

Only two of the Larne trawlers had yet been fitted with W/T, *Nellie Braddock* and *Ceresia*. The minesweeping trawlers were permanently transferred to Buncrana later in the month. The SNO at Buncrana was Captain Frederick Crooke, who was succeeded by Captain Herbert Chatterton in mid-1915. Berehaven and Killybegs were also captains' appointments, with Captain Robert H Travers at the former and Captain Arthur P James at the latter. Travers was succeeded by Commander Odienne Coates. Killybegs

Above: A postcard by Fergus O'Connor of warships in Bantry Bay. *(Adrian Healy Collection)*

Below: Warships in Bantry Bay. *(Lawrence Collection)*

The SS *Hibernia* at Holyhead in 1907. *(Anglesey Môn Information)*

in Co Donegal was described by Black's Guide as, 'a fishing village, 'nicely situated on the edge of its large and natural land-locked harbour, almost circular in shape, in which ships of large size can anchor at any state of the tide'. The closest town to Berehaven harbour in Bantry Bay was Castletownbere, 'the only town of any consequence in the extreme western part of Co Cork.' It was also noted that, 'Once a year the Channel Fleet holds here its autumn manoeuvres.'

Galway was also designated as a naval base, as noted in the *Connacht Tribune* of 23rd January 1915, with the armed trawlers *Saxon* and *Guillemot* arriving in Galway docks in February. As the flotilla expanded, duties would include patrols by night searching for mines, minesweeping daily in Galway Bay, examination vessel duties regarding ships entering the bay and patrols as far south as Tearaght Island off the Dingle Peninsula and as far north as Eagle Island in Co Mayo. Other duties were noted, including taking a Coastwatchers' hut from Westport to Inishboffin and the band and officers of the Connaught Rangers to Kilronan on the Aran Islands for a recruiting meeting.

The largest ship to be based in Larne in 1915 was the former London and North Western Railway, Kingstown to Holyhead mail steamer, TSS *Hibernia*, under the command of Captain Edward Tanner, which was requisitioned by the Admiralty and renamed HMS *Tara*, Tanner remaining as a Lieutenant RNR under Captain Rupert Gwatkin-Williams. It was decided that operational control away from harbour would be vested in *Tara's* CO, 'assigning stations and movements to patrol vessels at sea according to varying requirements, wind and weather conditions and the number of vessels available.

Gwatkin-Williams later wrote about his experience in *Tara*:

'It was there, in the old North Channel, that we were destined to spend our first year of the war, a humble but, I believe, useful servant of the State. A very happy year it was to me; it was one of no great excitement, but it was nevertheless full of emotion, of thrills, of hopes and of fears. In that first year we steamed 60,000 miles in the narrow waters; during the first few months we actually did thirteen days at sea out of every fortnight, and the fourteenth was spent coaling in harbour, either at Larne or Campbeltown. We never had a breakdown, and we faced out every kind of weather. In fact, I believe we held the record for both mileage covered and number of days at sea for the whole British Navy. My personal duties at first were very light, for, according to the Admiralty instructions given us, the ship's old captain [Lieutenant Tanner] had sole control over the handling of the ship. My own work was really of an advisory nature. I was a "go between" the Company and the Admiralty, an interpreter of official orders, a link when working with other naval units. But on emergency Captain Tanner was bound to obey my orders on my giving them to him in writing. As may be imagined, this was a most unsatisfactory arrangement for both of us, and it was only our mutual personal good will which rendered its working at all possible, and the system of dual control was not repeated in any other ships subsequently commissioned.'

Below: A newspaper report on the loss of HMS *Viknor*. *(Author's Collection)*

LOSS OF H.M.S. VIKNOR.

THREE PORT BOYS GO DOWN WITH THE SHIP.

ALL WROUGHT IN MESSRS RUSSELL'S DOCKYARD.

TWO MEMBERS OF ST JOHN'S CONGREGATION.

ONE MEMBER of the PARISH CHURCH.

HAD ALL BEEN AT THE SIEGE OF ANTWERP.

One of the greatest calamities which has befallen the Port in the present terrible war is the foundering of H.M.S. Viknor somewhere off the North of Ireland. Three gallant sons of the Port have gone down with that ship not so far, it may be, from the birthplace of two of them. Towards the end of December we published the

photo of the three boys. Their names are Joseph Duffy, whose widowed mother reside in Bouverie; Samuel Gourlay, who resided in Carnegie Park Gardens; and Samuel Mitchell, third and youngest son of Mr William Mitchell, residing at Ashgrove Buildings. All three lads were employed in Messrs Russell & Co.'s Dockyard. Samuel Mitchell, whose father is also employed in that yard, had almost completed his apprenticeship as a shipwright at the outbreak of the war. Joseph Duffy and Samuel Gourlay were members of St John's Church, while Samuel Mitchell was a member of the Parish Church, and is the first of the Roll of Honour in the vestibule of that church to lose his life in the war. His second brother is James Mitchell, formerly employed in the Clyde Shipbuilding and Engineering Company's works, and who is now stationed at Fort Matilda with the Clyde Royal Garrison Artillery.

Readers may remember that on the 16th of October we published the list of between thirty and forty boys of the Naval Reserve from Port-Glasgow who had come home on furlough after taking an active part in the siege of Antwerp. The names of the three lads were in that list. Accompanying their portraits in our issue of the 18th December were the following remarks: "They were all in the trenches during the historic siege of Antwerp, and saw the German shells flying over their heads towards the doomed city. They took part in assisting the Belgians to flee before the advance of the cruel Germans. Having reached the coast in safety after many trying experiences they were granted five days' leave. They reached the old Port along with their comrades on 14th October, and on the expiry of their furlough they were soon again rallying round the grand old British flag. They proceeded to Portsmouth and went under training in H.M.S. Excellent. They are now all A.B.'s, and have recently joined H.M.S. Viknor on the Tyne. They were not so very many miles away when the German cruisers came over in the fog and shelled the towns on the Yorkshire and Durham coast."

It is a month to-day since these gallant lads sailed away from the Tyne on some unknown destiny. They were buoyant with the spirit and hopefulness of youth. They wrote home saying they expected that their ship would call at some port once in three weeks, and that they would send word whenever they reached shore. But that ship was destined to make history repeat itself in so far as she proved herself to be the ship which never returned. There seems nothing more certain than H.M.S. Viknor is another victim of the mines cast over the seas by our cruel and dastardly enemies the Germans.

The following is the official intimation of the loss of the ship with all hands: "The Secretary to the Admiralty regrets to announce that the armed merchant vessel H.M.S. Viknor, which has been missing for some days, must now be accepted as lost, with all officers and men. The cause of her loss is uncertain, but as some bodies and wreckage have been washed ashore on the north coast of Ireland it is presumed that during the recent bad weather she either foundered or being carried out of her course struck a mine in the seas where the Germans are known to have laid them."

The Viknor was commissioned on December 12, 1914. No indication is given of the number of her crew, but the list of 21 officers, including the commander, appears in the Navy List published at the beginning of this month.

BODIES WASHED ASHORE.

Samuel Mitchell was a native of 'Derry,

Meanwhile, the year of 1915 had begun with the loss of two naval vessels off the north coast of Ireland. The armed merchant cruiser HMS *Viknor*, Commander Ernest Ballantyne, departed from Londonderry on 13th January and headed west towards the Atlantic for her patrol area. That same day she sank with all hands in heavy seas off Tory Island. She sent no distress signal.

HMS *Viknor*. *(Author's Collection)*

Wreckage and many bodies were washed ashore on the north coast of Ireland. Some of the crew were buried on Rathlin Island and Ballycastle in Co Antrim. Within a few weeks, on 3rd February, another armed merchant cruiser also from 10th Cruiser Squadron was lost, HMS *Clan MacNaughton*, Commander Robert Jeffreys, again with all hands, to the west of the Hebrides. It is possible that both were sunk by mines.

HMS *Clan MacNaughton* *(Author's Collection)*

The first 35 net drifters had arrived in Larne towards the end of February. The CO of the drifter flotilla was Commander George Ward of the Armed Yacht *Clementina*. On 22nd nets were laid over a five mile stretch of sea by 10 drifters. To begin with it was not a great success as 31 nets and 20 buoys were lost, 'from a cause not yet ascertained' to which the report's writer, Sub Lieutenant Robinson, added, 'nets cannot be used efficiently except in moderately fair weather, rough seas bring the dangers of losing the nets and damaging boats.' By the end of the month 65 were being fitted out with nets and buoys and over the next few months there was a ongoing and very intensive process of trial and error, testing the best way of operating with the nets and developing remedial measures to iron out snags.

As Chatterton wrote, 'No class of men in existence knew more about nets than these drifter men who had been handing herring nets since they were boys. So much had to be learned by experiment because it was a new method of which no data existed for guidance. It afforded a fine opportunity for this paricular class of fishermen to show their enthusiasm and originality, some of the important improvements in this connection being their own invention.' They also had to contend with, 'strong tides, dangerous overfalls, heavy seas, rain and mist, with no safe anchorage adjacent, but gaunt wild cliffs and barely two months of the year of tolerable conditions – these little vessels could not be an infallible protection.'

It was recommended that each section of 10 drifters should be under the command of a Sub Lieutenant and that one armed trawler should also be allocated to each four day shift at sea. The Admiralty suggested that accommodation for officers could be provided by

Drifters on patrol. *(Author's Collection)*

Paying out an anti-submarine net and floats from a drifter. *(Author's Collection)*

converting part of the drifter's fish hold or by adding a cabin to the upper deck. This work was carried out in Belfast. Much other work on the drifters and trawlers was undertaken by the Larne Shipbuilding Company (which had been established in 1878 near the Olderfleet Castle and would survive until 1922). It was noted that the company was under a certain amount of strain and needed to improve its output:

> 'The lack of skilled labour is much felt, and defects accumulate both to deck, hull, and engine room. In spite of eight Drifters having been sent to Belfast to dock and refit, the Larne Shipbuilding Co has been unable to cope with the work; although they have a yard which is capable of a good output and considerable development if skilled labour were procurable. The position of this yard makes it the more desirable that it should be brought up to the standard required, for it has the advantage of being near at hand, so that Drifters should be able to be repaired under the direct observation of the Commander-in-Charge, and at the same time their crews be used for the fitting of nets and the many duties for which they are required at the Base.'

The company was taken in hand, 'extensive additions were made to the facilities and a great many extra hands were engaged.' To resume the narrative of Captain Gwatkin-Williams:

> 'Gradually, however, my own sphere of activity became enlarged, as in course of time a quite respectable flotilla, including four destroyers, came into being and hinged upon the *Tara*. We were the only ship in these waters which had efficient wireless, the only one that could keep the sea on all occasions, and I was, moreover, the Senior Officer afloat in those waters. I posed as a demi-admiral and benevolent autocrat. The days spread themselves out into weeks, the weeks into months, but winter or summer, the beauty of that wild North Channel gradually sank into me – I no longer wished to leave it. The great billowing clouds for ever drifting from the westward like some painted Armada in full sail, the mist, the sunshine, the storms lashing the great basalt columns of Fair Head, the tumbling tide-ripped waters of that narrow straits – I know no more beautiful spot in the world – I grew to love them all; it was home to me. The ship was for ever busy with boarding work, and at night we scared many an inoffending tramp steamer by switching our search-light on to her at close quarters in the darkness; this done, we would vanish, zig-zagging rapidly into the blackness of the night, lest a lurking U-boat had marked us.'

Chapter 2

The U-boat War – the First Phase

After the repulse of the initial German advance and the establishment of a system of trenches running from Belgium to the Swiss border, the war in France had come to something of a stalemate. A decision was made to try and break the deadlock by the application of a change of tactics at sea. It is believed that the first U-boat to penetrate into the Irish Sea – as far north as Barrow-in-Furness – was U-21, Kapitänleutnant Otto Hersing, who was 28 years old and the son of a Strasbourg University professor, which set out from Germany on 23rd January 1915.

He came through the Dover Straits by night, proceeded the whole length of the English Channel, turned north and passed the Welsh coast into the Irish Sea. On 29th January, having arrived off Walney Island, the U-boat opened fire on the airship shed. On the following day it sank by boarding and placing explosive charges on three steamers in Liverpool Bay – an Admiralty collier and two cargo vessels bound for Belfast – *Ben Cruachan* (which was loaded with coal bound for the Grand Fleet at Scapa Flow), *Kilcoan* and *Linda Blanche*. It would appear that Hersing acted within the rules of war. For example, the Acting Master of the *Ben Cruachan*, David Heggie, later remarked, 'The German Officers were very gentlemanly, I give them credit for that.' Closing on the *Kilcoan* (a Howden Brothers steamer from Larne) with the deck gun trained, the crew was ordered to abandon ship and come alongside. The Master, James Meneely, was sent back for the ship's papers, accompanied by

U-18 a typical pre-1914 U-boat marginally smaller than U-21. It was later sunk by HMS *Garry*. *(Author's Collection)*

Kapitänleutnant Otto Hersing of U-21. *(Author's Collection)*

Survivors from a ship sunk by a German U-boat. *(Author's Collection)*

four armed seamen with an explosive device and two yards of fuse, a charge was set port-side amidships and lit. After making provision for the safety of the crew, Hersing returned and fired shells into the hull to speed her sinking. U-21 made a safe return home to Germany. Chatterton, later wrote, 'A very large portion of this successful, enterprising spirit which was actuating the German U-boat service was owing to Hersing. His cruises were certainly extraordinarily daring, and showed considerable endurance. In other words, they afforded invaluable data from which to deduce the theory that much more could be expected of submarines, provided they were multiplied in numbers and built of improved designs.' The American writer, Lowell Thomas, met Hersing after the war:

> 'I found the celebrated under-sea raider to be a tall, dark, slender man, with all the dignified and hospitable courtesy of a German rural proprietor. The pictures I had seen of him – war-time pictures – showed a striking-looking young chap with a keen, hawk-like face – a devil of a fellow to all seeming. He looked much older nearly ten years after the Armistice. He told his callers that he was troubled by rheumatism, a malady that submarine men commonly contracted from the continual dampness of boats. When we asked him what he was doing, he replied: "I grow fine potatoes."'

On 4th February a communiqué was issued by the Imperial German Admiralty which declared:

A fine study of the SS *Kilcoan* in Larne Lough by Norman Whitla. *(Courtesy of Larne Museum & Arts Centre)*

> 'All the waters surrounding Great Britain and Ireland, including the whole of the English Channel, are hereby declared to be a war zone. From 18th February onwards every enemy merchant vessel found within this war zone will be destroyed without it always being possible to avoid danger to the crews and passengers. Neutral ships will also be exposed to danger in the war zone, as, in view of the misuse of neutral flags ordered on 31st January by the British Government, and owing to unforeseen incidents to which naval warfare is liable, it is impossible to avoid attacks being made on neutral ships in mistake for those of the enemy.'

This declaration opened the first phase of what was to become known as unrestricted submarine warfare. Restricted submarine warfare had meant that, in compliance with the Hague Conventions of 1899 and 1907, the U-boat would surface, warn its intended victim, give the crew time to abandon ship and then sink it. Neutral cargo ships and all passenger liners, probably even Allied ones, would be spared.

In order not only to prosecute the war and to supply its troops with food and munitions but also to survive on the home front, Britain relied heavily on seaborne trade. If Germany could break or even seriously disrupt the flow of merchant vessels then Britain's ability to have waged war or indeed feed its population would have been rendered either difficult or impossible. Among the Royal Navy's many tasks were protecting trade, guarding the sea lanes and making safe the approaches to the major ports. Moreover, the German claim that unrestricted submarine warfare was justifiable was a response to the Royal Navy's blockade of German shipping and any neutral vessels deemed to be carrying supplies useful to Germany.

The tenuous nature of this claim can be challenged by considering the words of Admiral Reinhard Scheer, the Commander-in-Chief of the High Seas Fleet from January 1916 to August 1918, 'A U-boat cannot spare the crews of steamers, but must send them to the bottom with their ships. The gravity of the situation demanded that we free ourselves of all scruples.' However, the declaration was endorsed with enthusiasm by the commander from 1915 to 1916, Admiral Hugo von Pohl.

Below left: A neutral Swedish vessel photographed during the First World War. *(Jack McCleery)*

Below centre: Admiral Reinhard Scheer. *(Author's Collection)*

Below right: Admiral Hugo von Pohl. *(Creative Commons Share Alike 3.0)*

Kapitänleutnant Walther Schweiger. *(Author's Collection)*

On 10th February U-30, Korvettenkapitän Erich von Rosenberg-Gruszczyski, left Heligoland and sank two ships in the Irish Sea on 20th, SS *Downshire* out of Dundrum to Manchester, by means of explosive charges and SS *Cambank*, in this case by torpedo, without warning, causing four deaths.

U-20, commanded by Kapitänleutnant Walther Schweiger, 'a 32 year-old bachelor from an old and respected Berlin family – calm, correct and coldly efficient' and U-27, Kapitänleutnant Bernd Wegener, were both operational in the Irish Sea by the end of February, having taken the southern and northern routes respectively. On 13th March Wegener sank the SS *Hartdale* off the Co Down coast, seven miles south of the entrance to Belfast Lough. The first time a U-boat appeared off the west of Ireland was in the middle of March, when the SS *Atalanta* was attacked and set on fire off the coast of Connemara by U-29, Kapitänleutnant Otto Weddigen. Only four days later, on 18th March, this U-boat would perish with all hands after being rammed by HMS *Dreadnought*, Captain William Alderson, in the North Sea.

U-29 approaches a merchant ship. *(Author's Collection)*

On the south-east coast the first sinking was on 28th March, when the Elder Dempster liner, SS *Falaba* was sunk by U-28, Kapitänleutnant Freiherr Georg-Günther von Forstner, in St George's Channel. *Falaba* was hailed through a megaphone to, 'take to the boats, as we are going to sink the ship in five minutes.' This was at noon, and ten minutes later the submarine fired a torpedo from a distance of about 100 yards apparently in reaction to *Falaba* sending wireless messages and distress rockets for help. *Falaba* took a list to starboard and sank in eight minutes. The steamer carried a crew of 95 and 147 passengers, including seven women, a total of 242 persons; and it was quite impossible to transfer this number of people to the boats in the bare twenty minutes that elapsed between the U-boat's warning and the sinking of the ship. While the boats were being launched at top speed, the falls of one boat slipped, the falls of another jammed, some boats were dashed against the side of the ship, and one was seriously injured by the explosion of the torpedo. The result was that 104 lives were lost, one of whom was a US citizen, Leon Thresher, a mining engineer headed towards the Gold Coast in Africa. 138 lives were saved. The American press denounced the sinking as a massacre and an act of piracy, but the US Administration took no action on the matter at this time.

A bows-on view of HMS *Dreadnought* as would have been seen from U-29. *(Author's Collection)*

This was the nature of the problem facing the Admiralty

under the political direction of the First Lord, Winston Churchill. However, he recorded in his book, *The Great War*, the Admiralty was not too concerned at that stage. The Germans had moved too soon, there were simply not enough U-boats available in 1915 to make unrestricted submarine warfare any more than a considerable nuisance rather than a major threat. Only 23 were available, with many more not yet ready for service or still being constructed. They compounded this error with a major miscalculation. On a clear, windless day, at 2.10 pm on 7th May 1915, off the Irish coast, eight miles south west of the Old Head of Kinsale, the Cunard Line's RMS *Lusitania*, commanded by the gruff, experienced seaman, Captain William Turner, was hit by a single torpedo fired at a range of 700 yards by Schweiger's U-20 (which had previously sunk the schooner *Earl of Lathom* and two steamers, *Candidate* and *Centurion*, off Waterford). The great ship was destroyed in a second, sudden, violent detonation (it was carrying a not wholly declared cargo of 5000 3.3-inch shrapnel shells without their explosive charge, 4.2 million live rifle cartridges and 46 tons of volatile aluminium powder), sinking in just 18 minutes, which resulted in the

A photo taken of RMS *Lusitania* on her last voyage to America in April 1915. *(Martin & Lindy Lovegrove Collection)*

Above: The tug *Flying Fish* went to the assistance of the *Lusitania*. *(Author's Collection)*

RMS *Lusitania (Martin & Lindy Lovegrove Collection)*

Wanderer, with the stricken *Lusitania* in the background. *(Courtesy Roy Baker, Curator of the Leece Museum, Isle of Man)*

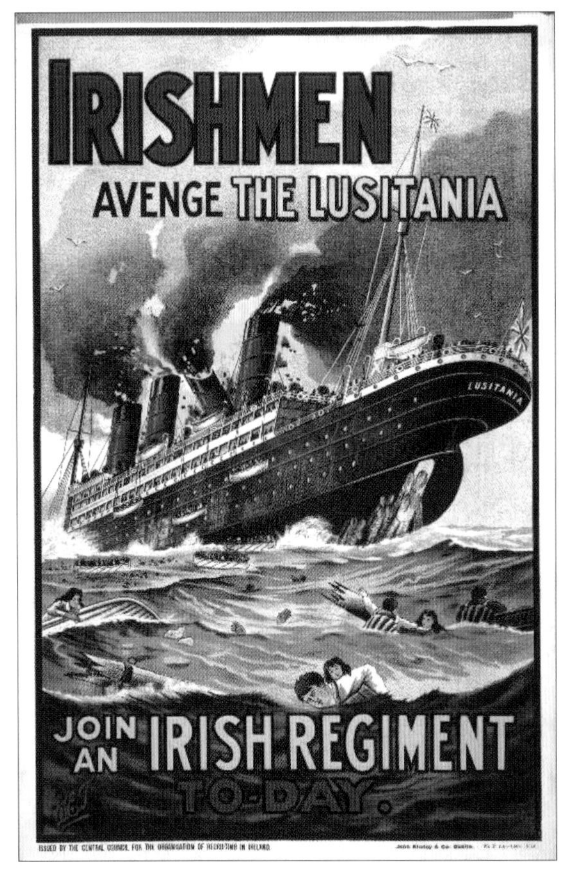

A recruiting poster following the sinking of the *Lusitania*. *(Library of Congress)*

death of nearly 1200 of the 1959 souls on board, despite the best efforts of armed trawlers and drifters from Queenstown, *Wanderer*, a two-masted lugger out of Peel, which had been fishing for mackerel, the tug *Flying Fish* and the RNLI lifeboat from Courtmacsheery. Over 100 of the dead were American citizens, one of whom was the multi-millionaire Alfred G Vanderbilt.

Former President Theodore Roosevelt denounced the sinking as, 'piracy on a vaster scale than any old-time pirate ever practised.' It is hard to balance such an act with the description of Schweiger's character as given by Leutnant Rudolph Zentner:

'She was a jolly boat, the U-20, and a kindly boat – and she sank the *Lusitania*. If you want a good and pleasant boat you must have a good and pleasant skipper. Kapitänleutnant Schwieger was one of the few U-boat officers who was in the submarine service when the war began. He was one of the ablest officers we had and a recognized expert on submarine matters – one of the few commanders who were consulted by Grand Admiral von Tirpitz and on whose advice Von Tirpitz relied. The records credit him with having sunk 190,000 tons of Allied shipping. He was well educated and had in the highest degree the gifts of poise and urbane courtesy. He was tall, broad-shouldered, and of a distinguished bearing, with well-cut features, blue eyes, and blond hair – a particularly fine-looking fellow. He was the soul of kindness toward the officers and men under him. His temperament was joyous and his talk full of gaiety and pointed wit. He had the gifts to command both respect and liking and was a general favourite in the German Navy.'

The situation was further aggravated on 19th August by the sinking of the White Star passenger liner, SS *Arabic* by U-24, Kapitänleutnant Rudolf Schneider, again off the southern coast of Ireland, with further American loss of life.

U-26 was of the same class as U-24. *(Author's Collection)*

Such was the outcry that the Germans prudently scaled down the unrestricted nature of the submarine campaign in order to avoid incurring greater wrath from the USA. It also highlighted the importance of having auxiliary patrols off the south-west coast of Ireland, giving the Irish command an importance which it had never previously possessed.

Queenstown

In August 1914 Queenstown had been, in the words of Lieutenant Commander Gordon Campbell, who, in October 1915, was the captain of the Q-ship HMS *Farnborough* at Queenstown (hence the 'Q'):

'...a small naval base which became of great importance later on, and from having a Vice Admiral in command it became the appointment of a Commander-in-Chief [Vice Admiral Sir Lewis Bayly succeeded Vice Admiral Sir George Coke as 'Senior

Queenstown. Admiralty House may be seen on the skyline to the right of the picture. *(Library of Congress)*

Officer on the Coast of Ireland' in July 1915 and remained in post until 1919, having been promoted to full admiral in 1917 and assuming the much more appropriate title of C-in-C Western Approaches]. There was a small but very efficient dockyard there on Haulbowline Island, which lay opposite the town of Queenstown [now Cobh], where the C-in-C had his residence and headquarters at the top of a hill and so had a good view over the harbour. It was the most suitable place for our operations, as it was easy to get to any of the many trade routes which approach the British Isles. It often struck me that Queenstown would have been an excellent place to have had a sort of Admiralissimo of all the approaches to the British Isles from the westward and southward, Plymouth, Milford, etc., being sub-bases. Whitehall is too far away, and the sea air doesn't penetrate so far.'

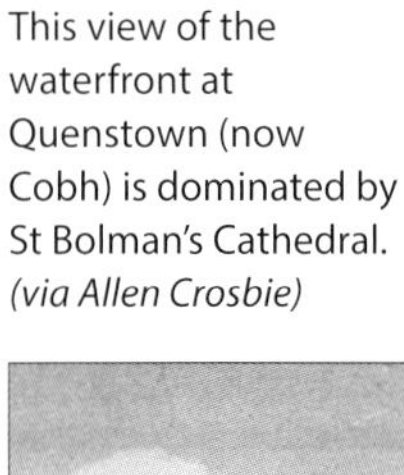

This view of the waterfront at Quenstown (now Cobh) is dominated by St Bolman's Cathedral. *(via Allen Crosbie)*

Above: The Esplanade, Cobh, formerly Queenstown. *(via Allen Crosbie)*

A flavour of life at Queenstown was given to the author by local historian, Allen Crosbie:

'In the Edwardian period Queenstown was prospering, as never before or since. It was the terminus for the main railway line from Dublin. The transatlantic liners called going and coming to deliver and pick up passengers and mail, and the harbour was full of sailing ships inward bound from China, India, Australia and New Zealand. Life for the professional classes in Queenstown was more leisured and pleasant than at any time since. The Cork Opera House was flourishing and provided both light and classical operas, and symphony orchestras came there for a season. Queenstown had its own Musical Hall. There was a pack of beagles on the island and the major hunts on the mainland, and a lot

John Gilbert, Patrick's Quay, Cork 1886. Watercolour on paper 21 x 31 cm. *(Crawford Art Gallery Cork)*

The armoured cruiser HMS *Sutlej* of the 11th Cruiser Squadron. *(Author's Collection)*

> of good shooting on the marsh land round about. Only the wealthy would have possessed guns. Queenstown accommodated Cork business men, the families of the officers on the Royal Naval ships and the dockyard, and most of the families of the Army officers stationed on Spike Island. The Admiral was the most important local dignitary.'

In 1914 the 11th Cruiser Squadron West of Ireland Coast Patrol was based at Queenstown on trade route protection duties, to which could be added four 1886-vintage Torpedo Boats, HMTB 050, HMTB 052, HMTB 055 and HMTB 058. The depot and receiving ship at Queenstown was HMS *Colleen*, the former corvette HMS *Royalist*, hulked in 1900 and renamed in 1913. Later in the war the famous Irish Antarctic explorer, Tom Crean, would serve in *Colleen*.

Vice Admiral Lewis Bayly RN. *(Naval History and Heritage Command (NHHC))*

The elderly and ineffective cruisers were soon dispatched to the far north of Scotland.

In July 1915 Bayly was serving as President of the Royal Naval College, Greenwich when the First Lord of the Admiralty, Arthur Balfour, sent for him and invited him to accept the Queenstown command, pointing out that securing the safe passage of ships from the USA to Britain through the danger zone to the north and south of Ireland, with munitions of war, guns, oil, machinery and foodstuffs made the post a very important appointment.

His reply was characteristically brief and to the point. He took pencil and chart, and, with Queenstown at the centre, drew a circle which included Milford Haven, the Scilly Isles

HMS *Pathfinder* a Scout Cruiser similar to *Adventure*. *(Author's Collection)*

(transferred to Devonport Command in March 1916), the Irish Sea, the North Channel and all the waters around Ireland. He then said, 'If you can give me command in that circle and if you give me a fast cruiser, I will accept.' Within a few days Bayly boarded the cruiser, HMS *Adventure*, arriving at his new command on 22nd July. It would remain on half hour's notice by day and ten hours' by night to allow the Admiral to visit patrol ships at sea, other bases in his command or vessels in distress. Privately he hoped for the opportunity to fight a German surface raider but none came calling.

Campbell, who later became his Flag Captain, in HMS *Active* in November 1917, had this to say for Bayly, 'He was the finest C-in-C I have ever served under and a man I had the greatest respect and affection for.' Others would describe him as brusque with little, if any, small talk, an exacting disciplinarian, authoritarian and a hard driver. At heart he was a thoughtful, intelligent and well-read officer, with a great streak of kindness that inspired loyalty and affection. Indeed Vice Admiral William Sims USN would later write of his friend, 'This weather-beaten sailor had a great love for flowers, for children, for animals, for pictures, and for books; he was deeply read in general literature, in history, and in science, and that he had a knowledge of their own country and its institutions which many of our own officers did not possess.'

Right: Vice Admiral William S Sims USN. *(NHHC)*

Far right: Captain Gordon Campbell VC RN. *(Author's Collection)*

Campbell himself was described thus by Admiral Sims, who admired his bravery, intelligence and skill, 'He was a short, rather thick-set, phlegmatic Englishman, somewhat non-committal in his bearing; until he knew a man well, his conversation consisted of a few monosyllables, and even on closer acquaintance his stolidity and reticence, especially in the matter of his own exploits, did not entirely disappear.'

Under Bayly's leadership Queenstown would become a vital base in the war against the U-boats. Chatterton praises Bayly highly:

> 'He thoroughly reorganized the patrol forces. There were all too few craft available, but a flotilla of sloops [newly built as fleet minesweepers, with single screws, capable of 16–17 knots and with an operating range of about 2000 miles; but actually used for patrol and sweeping as required. They began arriving not long after the Vice Admiral and were about the size of a contemporary cross-channel packet steamer] began to guard the trade routes where they converged on the Irish coast; armed trawlers patrolled inside of the sloops; net-drifters now shot their nets in likely bays, waited and hoped; whilst motor-boats with RNVR officers patrolled the numerous bays, creeks, and inlets which are a feature of south-west Ireland, and might be used by lurking submarines. The motor-boats also had to keep an eye on this lonely coast to prevent the landing of arms and supervise the local fishing craft.'

He also had four steam yachts, *Aster II*, *Beryl*, *Greta* and *Pioneer II*. Chatterton describes Bayly's burden, 'For the Admiral all this control and direction; the strategy of patrols; the responsibility for the safety of valuable heavy-laden steamers passing through his area; the supervision of Haulbowline Dockyard; the rapidly growing numbers of sloops, yachts,

One of the patrol ships to arrive at Queenstown was the 24 Class sloop HMS *Flying Fox*. *(NMRN)*

trawlers and drifters; the immense amount of official correspondence and paper work; the hourly incoming and outgoing telegrams and wireless signals, this was a task so heavy, so thankless at times, that in those early days it might by some have seemed unbearable.'

He also described the work of Haulbowline Dockyard:

> 'To keep these little cruisers [the sloops] at sea for five days with only a couple of days in harbour, yet maintain personnel, engines, hulls, and fittings in a high state of fighting efficiency, was no small achievement of organisation. During those two days the ship had to go alongside a collier and laboriously fill bunkers with so many tons of coal; food and other stores had to be fetched aboard, and any defect made good, whether arising from fair wear and tear, unfair seas, or any other cause. To this end the dockyard, its officials and men, its facilities and store houses were kept ever in readiness to deal with any emergency and really help a captain by answering his needs without tying him up in a muddle of red tape.'

The same may be said for all the other Irish bases, depending upon the level of facilities offered at each.

Bayly commented, 'The sloops were at first rather an anxiety; they were a new type of ship, mostly commanded by youngsters who had never commanded anything before, and gales and heavy seas were frequent. But the anxiety soon passed; they were excellent sea boats,

Above: A Queenstown sloop putting to sea. *(Author's Collection)*

Right: HMY *Beryl*. *(Author's Collection)*

and the youngsters who commanded them were splendid. They faced the gales, chanced the fog, and were always ready to face any emergency, no matter how hard the work or how heavy the sea.'

Bayly's tact and naval directness also won the co-operation of the Bishop of Cloyne, the Most Reverend Robert Browne, whose cathedral dominated the skyline close to Admiralty House. The Admiral later wrote, 'I pointed out to him that he was really the most important personage in Queenstown, and that things would be better in every way if we worked together. The Bishop entirely agreed, and from that moment began a friendship between us that has lasted for many years.' Bayly facilitated the safe transport of 42 bells for the cathedral from Liverpool and also successfully interceded, at the Bishop's request, in the case of a 'foolish Irish lad aged nineteen' condemned to death by a court martial in the aftermath of the Easter Rising. In return the Bishop used his influence to smooth out any difficulties in relationships with his flock.

Larne

By the beginning of 1915 no less than 21 Auxiliary Patrol Areas had been created embracing the entire coastline of the British Isles. In early December 1914 a scheme had been worked out for creating these Patrol Areas. It was calculated that 74 steam yachts and 462 trawlers and drifters would be required.

In Larne, Auxiliary Patrol Area XVII, the Senior Naval Officer was Temporary Captain RNR Charles J Barlow, who was actually 66 years old and a retired Admiral, who had returned to service in 1914 in command of HMY *Valiant*, which had been owned by Lord Pirrie, the Chairman of Harland & Wolff. Barlow had been given a 'flying squadron' of six large, armed yachts, *Valiant*, *Jeanette*, *Marynthea*, *Medusa*, *Narcissus* and *Sapphire* to supplement the craft already based at Larne. The war was coming closer to Larne; on 11th March 1915 HMS *Tara* assisted in recovering the survivors from the Armed Merchant Cruiser, HMS *Bayano*, Commander Henry Carr, which was torpedoed and sunk by U-27, 10 miles SE by E of Corsewall Point. *Bayano* was one of several Elders & Fyffes 'banana boats' taken up into the 10th Cruiser Squadron.

Barlow was replaced in April by Rear Admiral the Hon Robert Francis Boyle, who was 15 years younger. His last duty at Larne was an inspection of the drifter net field in *Clementina*. The North Channel was being taken in hand. Here, as distinct from the English Channel, a modification was adopted, as these straits were both narrower and deeper than those separating England from France. Anti-submarine indicator nets were laid between Rathlin Island and the Scottish mainland, as well as at other points in the

HMS *Bayano*. *(Author's Collection)*

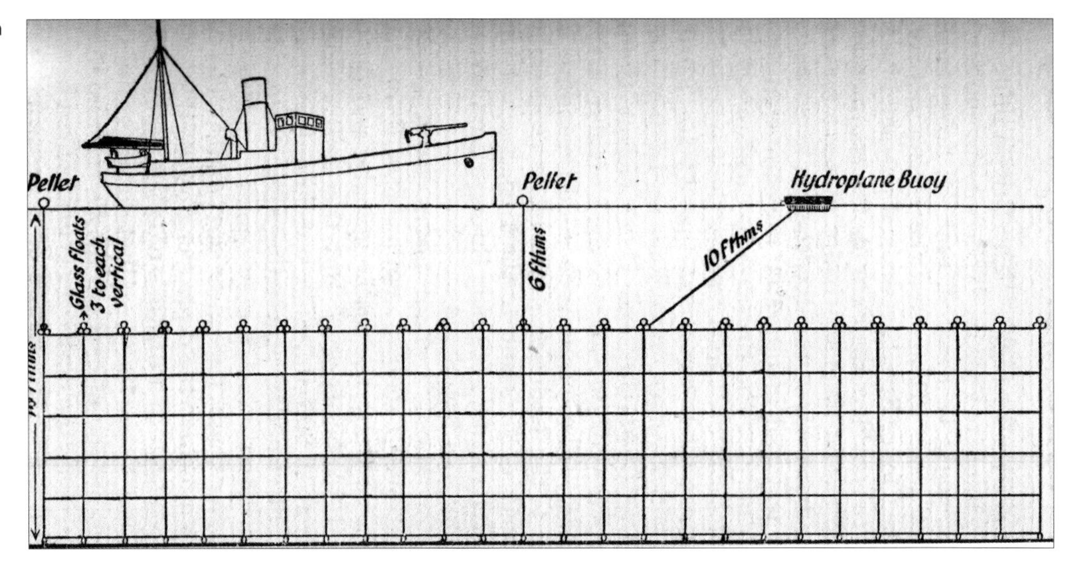

Diagram of a drifter with an anti-submarine net. *(Author's Collection)*

North Channel. These were maintained by the men of eventually more than 130 craft based in Larne, which would play a role out of all proportion to their size and armament. As the North Channel was too deep to net to the bottom, the object was to make the submarines use up their batteries by keeping them submerged for as long as possible. Two lines of nets were laid 20 miles apart, 'a light flexible curtain of thin steel wire woven into six or ten foot meshes and supplied in lengths of 200 yards. These were laid clipped together. The submergence of the glass buoys on which the nets were hung or the automatic ignition of a calcium light betrayed immediately the presence of a submarine.' Between these nets were four or five lines of net drifters, supported by patrols. Day and night the little steamers kept under way, continuously towing their nets. In addition to the system of nets there were surface patrols for five miles at each end, which meant that to pass through, a U-boat had to stay submerged for 30 miles. A patrolled passage three miles wide was left open for shipping on the Irish side. Such was the plan but naturally it encountered many 'difficulties and disappointments' when put into practice. A report noted, 'Four drifters have been fitted out with sets of ten 60 foot nets fitted with glass balls and wire foot ropes. The nets appear to work most satisfactorily, but Indicator buoys are a failure. Nets have been repaired and fitted at the average rate of five per day, and the men are steadily improving in workmanship. The first mine dropping shoot was completed on Tuesday, April 7th.'

Leutnant zur See Rudolph Zentner. *(Author's Collection)*

Leutnant Rudolph Zentner described what it was like in a submarine that had become entangled in a net:

'We were running submerged. Through the speaking tube came a shout from the commander: "Two buoys sighted.

Keep exact depth," he ordered. Later on I should have known exactly what they meant, but then they seemed a bit peculiar but nothing more. Suddenly there was a peculiar racket. It sounded as if huge chains were banging against the boat and were being dragged over it. The men at the diving rudders shouted to me that the apparatus was out of control. A glance at the gauges showed me that our speed had slowed down and that we were sinking. The boat turned this way and that, lurching and staggering drunkenly. She continued to sink and presently hit bottom with a bump. We were in a hundred feet of water. I leaped up the ladder and looked out through the window of the conning tower. All I could see was a maze of meshes and chains and links. Now we knew the meaning of those buoys. They were supporting a net. We had run into the net and now were entangled in it. Later on such a net would have been hung with bombs, like tomatoes on a vine. Thank heaven, they had hung none on the net we rammed! But meanwhile we were caught. It was a new situation and seemingly a hopeless one. We were sure we were caught fast and could never get out of those deadly meshes. You can bet there was no laughing and singing on board now. Each man thought of his home in Germany and how he would never see it again. "Reverse engines," Kapitanleutnant Schwieger commanded. The only thing we could think of was to try to back out. There was a great straining and cracking and clanking, and then we heard a familiar whirr – the propellers of destroyers. They no doubt had been lurking in the distance in such a way that they could observe a disturbance of the net – telling of a big fish. Now they were coming to see if they could make matters worse for us. Luckily, they did not have depth bombs in those days, or we should have been done for. The gauges were the whole world to us now. I had never gazed at anything so eagerly before. Yes, we were backing. With a ripping and rending we were tearing our way out of the net. We were clear, and away we went. All that remained to worry us was the sound of propellers. It followed us. The destroyers were keeping right above us. We dodged to right and to left, and still that accursed sound. You can easily tell a destroyer from another ship by the sound of the propeller, which in a destroyer has a much higher note – a shrill, angry buzz. Our periscope was down, but still something was giving us away on the surface, for those destroyers kept after us, no matter how we went. They were waiting for us to emerge, to shoot at us or ram us. We couldn't guess what the trouble was, but merely kept on going, trying in vain to lose those persistent hounds that were on our trail. This kept up hour after hour. We ran blindly under water, keeping as deep as we could. We didn't know much about where we were going. Any attempt to rise to periscope depth and take a look through the asparagus [periscope], and we should probably have been rammed.'

The crew of a drifter laying out an anti-submarine net. *(Author's Collection)*

C Class 30-knot 'Turtle-back' destroyer similar to HMS *Thorn*. *(Author's Collection)*

Above right: HMS *Dee* *(Author's Collection)*

On 15th April all available craft were dispatched to Loch Ryan to lay a net field and carry out a search, as it had been reported to Larne that two 'German Naval Officers' had been seen on the shore. Nothing was found. By April 1915 the North Channel Patrol included the rather elderly destroyers HMS *Dee*, HMS *Dove*, HMS *Garry* and HMS *Thorn*. *Dove* and *Thorn* had curved 'turtle-back' forecastles and, while speedy for their age, had poorer sea-keeping qualities than the slightly younger River Class pair, built with high forecastles. Two of the TBDs, *Dee* and *Dove* were often on detached duty out of Liverpool. *Garry* had already accounted for one U-boat, having rammed U-18, Kapitänleutnant Heinrich von Hennig, in the Pentland Firth on 23rd November 1914.

Captain Gwatkin-Williams was not impressed by an encounter on the morning of 20th April:

> '7.30 am *Tara* turned back Norwegian *Thoisdal*, for Ardrossan, from prohibited area, and British SS *King's Lynn* from Drogheda to Newcastle-on-Tyne – boarded latter in Red Bay [near Cushendall], Captain professed ignorance of orders. Apparently a very stupid man, as we had to fire three rounds 6-pounder before he understood, and at dark the same evening he was just about to enter Rathlin Sound and was fired on again, although boarding officer had explained he was not permitted through after sunset – on latter occasion *King's Lynn* was not actually contravening regulations, but was just about to do so.'

The 30-knot 'Turtle Back' destroyer HMS *Dove*. *(Martin & Lindy Lovegrove Collection)*

The River Class destroyer HMS *Garry*. *(Martin & Lindy Lovegrove Collection)*

A new depot ship arrived in Larne in May, the old cruiser HMS *Hermione*, and would remain there for the following six months. Interestingly *Hermione* had an aviation connection. This small 2nd class cruiser was intended in 1910 to act as sea-going depot ship for the ill-fated Naval Airship No 1, which was going to be named *Hermione* but is better remembered now by the nickname bestowed upon it – *Mayfly*, which it sadly did not, breaking its back while moored at Barrow in 1911.

The protected cruiser HMS *Hermione*. *(NMRN)*

Also in May another 50 net drifters arrived, allowing 10 sections of 10–11 vessels to be formed. The aim was to have sections continuously steaming round with a double line of nets, day and night. It was also reported that only seven out of 131 drifters were armed owing to a lack of guns. The drifters came from fishing fleets all around the coast of Great Britian, including Falmouth, Poole, Yarmouth, Lowestoft, Peterhead, Banff, Fraserborough and Aberdeen.

Support was also given to Kingstown, when three of the yachts, *Narcissus*, *Marynthea* and *Medusa*, were sent there on detachment. The harbour facilities at Kingstown were useful because of their size and the unimpeded access they gave to the Irish Sea. Dublin Port was further upstream on the River Liffey and would have necessitated a longer journey at reduced speed in and out. In 1915 the Admiralty began to extend its facilities, providing additional accommodation and munitions storage capacity, with dredging being proposed for the additional mooring of patrol boats. In time the entire Victoria Wharf area was requisitioned by the Admiralty and a first rate auxiliary naval base was established. According to Roy

The elaborate stern of HMS *Hermione*. *(Author's Collection)*

Stokes, 'The naval station had access to naval wireless, telephone, shipping, road and rail facilities and was situated in the heart of a town filled with a loyalist and patriotic fervour.' The Kingston to Holyhead route was important for the transport of troops to and from Ireland. The SNO at Kingston was Rear Admiral Evelyn Le Marchant, who had been appointed to the position when Kingstown was detached from Liverpool. He was succeeded at the end of the year by Captain Henry F Aplin and in the middle of 1916, Captain Herbert Chatterton. On 15th May 1915 the small Armed Trawler *Berkshire* from Unit 70 was sunk eight miles east of Cushendall following a collision with HMY *Valiant II*. All hands were saved. At the end of the month the Armed Yacht Squadron was sent from Larne to Milford Haven, leaving only *Tara*, *Garry*, *Thorn* and *Clementina* of the larger vessels. Unit 70 was temporarily broken up to free its trawlers for escort duties. For example, on 21st May *Ceresia* escorted SS *Irishman* from the entrance of Red Bay to Ailsa Craig, where *Irishman* was handed over to HMY *Adventuress* of the Clyde Patrol, while *Rose II* and *Goshawk* escorted SS *Vulture* from Ardrossan to Belfast. The final task for *Jeanette* and *Sapphire II* out of Larne was escorting the Belfast-built monitor, HMS *Admiral Farragut*, on trials of her 14-inch guns. (The ship was named *Admiral Farragut* in honour of the United States Admiral David Farragut, However as the United States was still neutral, the ship was hurriedly renamed HMS *M1* on 31st May 1915. She was then named HMS *General Abercrombie* on 19th June 1915, and then renamed HMS *Abercrombie* on 21st June 1915!) The trawlers *Vera*, *Roxano* and *Diver* escorted the monitor to Avonmouth, returning on 1st June. Sub Lieutenant Peat of *Vera* reported to the SNO Larne that, while on passage, he had been instructed by the senior officer on the *Farragut* to leave station and detonate a mine by rifle fire, which he did.

Eventually the six armed steam yachts of the 'flying squadron' were dispatched to the Mediterranean, because of an increase of submarine activity there and the consequent need for more sea-going craft. The total number of drifters in Larne had risen to 151 and Acting Commander Louis Brooke-Smith assumed the post of Commander-in-Charge Drifters. Twelve sections of 12 was regarded as the maximum possible for Larne. On 1st June six drifters were transferred to Buncrana, *Alfred*, *Baden Powell*, *Bartonia*, *David B Summers*, *Snowdrop* and *Spring Flower*. An incident occurred on 7th June. The newly-built RN submarine S-2, under the command of Lieutenant John de B Jessop, was proceding south from Greenock on the Clyde. Off the Mull of Galloway, near Portpatrick she was fired on an opening range of 4000 yards by the Larne-based trawlers *Ceresia*, *Rose II* and *Vera*.

In foggy weather the submarine had lost contact with its escort, the trawler *Kodama*. As the fog broke up into banks, the lookouts on the three trawlers had spotted a submarine's conning tower. Lieutenant Bayne stated in his report, 'We at once put on all our speed possible, and as soon as the gun bore, opened fire. After about five minutes [having closed to 3000 yards] we made out flags on the submarine and the escorting trawler

One of the three S-class submarines, later sold to the Italian Navy. *(Author's Collection)*

HM Trawler *Vera (Larne Museum & Arts Centre)*

stood towards us. We then understood that the submarine was British and at once steamed out to enquire if any damage had been done.' Fortunately no damage was done 'though the last shots came very near' and no lasting blame was attached to those involved, however, and a considerable amount of paperwork was generated for the COs, especially Lieutenant AG Bayne of *Ceresia*. During the summer the number of drifters at Larne settled down at just over 100 but still only eight were armed with 3-pounders. *Clementina*, which carried two 6-pounders, welcomed a new CO in July, Lieutenant Commander J Miller, replacing Captain TP Walker RNR, who had been a Vice Admiral RN. Unfortunately his tenure was to be brief, as at 9.55 am on 5th August, *Clementina* collided with SS *Adam Smith* off Torcor Point on the coast of North Antrim. She was badly holed and beached on the shoreline. Her guns and ammunition were salved, assisted by drifters from Larne, which also protected the wreck by anchoring nearby at night. The trawlers had to assume even more patrol duties. In the meantime, Captain Gwatkin-Williams made two important points to the Rear Admiral in his report of 2nd August:

> 'If available, another vessel of sea-going capacity should work with *Tara* in North Channel. The destroyers are of little use in bad weather, and as they are constantly being required for escort in other waters, they have not time to properly overhaul their engines when in harbour. The same applies to *Tara* herself, as she has been running continuously with all boilers since war started, with the exception of two very short refits. At present, I would submit the North Channel is at times very inadequately guarded, use having to be made of armed trawlers, which cannot be depended upon to give quick information.'

He followed this up with another expression of his concerns on 30th August:

> 'The amount of shipping passing through North Channel has been steadily increasing in volume for the last two months, it being mostly Scandinavian, Danish, and British and Russian vessels bound for Archangel. With the small number of patrol vessels generally available it cannot be satisfactorily overlooked. The armed trawler which generally has to do the duties of the Eastern Patrol, has no method of challenging war vessels which use that side at night. Hostile vessels could probably pass through as easily as British, and without being recognized as hostile. Even if fired upon, the probability is that the SNO in North Channel would not be aware what was taking place, since there is no method of communication with patrol trawlers at night at that distance. It is quite likely that rockets and flashes of guns would be unobserved, except in very clear weather.'

Inside a U-boat

Turning aside from the main narrative for a while, what was it like to serve in an Unterseeboote? This account of life in German U-boat was written by Oberleutnant zur See Johannes Spiess, First Watch Officer of the early kerosine powered U-9, commanded by Kapitänleutnant Otto Weddigen, who he described as, 'quiet, courteous, the very reverse of a martinet. You did not feel like a subordinate when you served under him, but rather like a younger comrade.' Spiess continued:

> 'Far forward in the pressure hull, which was cylindrical, was the forward torpedo room containing two torpedo tubes and two reserve torpedoes. Further astern was the Warrant Officers' compartment, which contained only two small bunks and was particularly wet and cold. Then came the CO's cabin, fitted with only a small bunk and clothes closet. Whenever a torpedo had to be loaded forward or the tube prepared for a shot, both these cabins had to be completely cleared out. Bunks and clothes cabinets then had to be moved into the adjacent officers' compartment, which was no light task owing to the lack of space in the latter. In order to live at all in the officers' compartment a certain degree of finesse was required. The Watch Officer's bunk was too

Captain Weddigen and the crew of U-9. *(Library of Congress)*

small to permit him to lie on his back. He was forced to lie on one side and then, being wedged between the bulkhead to the right and the clothes-press on the left, to hold fast against the movements of the boat in a seaway. The occupant of the berth could not sleep with his feet aft as there was an electric fuse-box in the way. At times the cover of this box sprang open and it was all too easy to cause a short circuit by touching this with the feet. Under the sleeping compartments, as well as through the entire forward part of the vessel, were the electric accumulators which served to supply current to the electric motors for submerged cruising. On the port side of the officers' compartment was the berth of the Chief Engineer, while the centre of the compartment served as a passageway through the boat. On each side was a small upholstered transom between which a folding table could be inserted. Two folding camp-chairs completed the furniture. While the CO, Watch Officer and Chief Engineer took their meals, men had to pass back and forth through the boat, and each time anyone passed the table had to be folded.

The hatch giving access to the periscope, hand wheels for pressure and valve gauges. *(Tyne & Wear Museum)*

Further aft, the crew space was separated from the officers' compartment by a watertight bulkhead with a round watertight door for passage. On one side of the crew's space a small electric range was supposed to serve for cooking – but the electric heating coil and the bake-oven short-circuited every time an attempt was made to use them. Meals were always prepared on deck! For this purpose we had a small paraffin stove such as was in common use on Norwegian fishing vessels. This had the particular advantage of being serviceable even in a high wind. The crew space had bunks for only a few of the crew – the rest slept in hammocks, when not on watch or on board the submarine mother-ship while in port. The living spaces were not cased with wood. Since the temperature inside the boat was considerably greater than the sea outside, moisture in the air condensed on the steel hull-plates; the condensation had a very disconcerting way of dropping on a sleeping face, with every movement of the vessel. Efforts were made to prevent this by covering the face with rain clothes or rubber sheets. It was in reality like a damp cellar.

The storage battery cells, which were located under the living spaces and filled with acid and distilled water, generated gas [hydrogen gas] on charge and discharge: this was drawn off through the ventilation system. Ventilation failure risked explosion, a catastrophe which occurred in several German boats. If sea water got into the battery cells, poisonous chlorine gas was generated. From a hygienic standpoint the

The engine room of a U-boat. *(Author's Collection)*

sleeping arrangements left much to be desired; one awoke in the morning with considerable mucus in the nostrils and a so-called "oil-head".

The central station was abaft the crew space, closed off by a bulkhead both forward and aft. Here was the gyro compass and also the depth rudder hand-operating gear with which the boat was kept at the required level. The bilge pumps, the blowers for clearing and filling the diving tanks – both electrically driven – as well as the air compressors were also here. In one small corner of this space stood a toilet screened by a curtain and, after seeing this arrangement, I understood why the officer I had relieved recommended the use of opium before all cruises which were to last over twelve hours.

Protective clothing as worn by U-boat crewmen on deck. *(Author's Collection)*

In the engine room were the four Körting paraffin [kerosine] engines which could be coupled in tandem, two on each propeller shaft. The air required by these engines was drawn in through the conning-tower hatch, while the exhaust was led overboard through a long demountable funnel. Astern of the gas engines were the two electric motors for submerged cruising. In the stem of the boat, right aft, was the after torpedo room with two stem torpedo tubes but without reserve torpedoes.

The conning tower is yet to be described. This was the battle station of the CO and the Watch Officer. Here were located the two periscopes, a platform for the Helmsman and the "diving piano" which consisted of twenty-four levers on each side controlling the valves for releasing air from the tanks. Near these were the indicator glasses and test cocks. Finally there was electrical controlling gear for depth steering, a depth indicator; voice pipes; and the electrical firing device for the torpedo tubes. Above the conning tower was a small bridge which was protected when cruising under conditions which did not require the boat to be in constant readiness for diving: a rubber strip was stretched along a series of stanchions screwed into the deck, reaching about as high as the chest. When in readiness for diving this was demounted, and there was a considerable danger of being washed overboard. The Officer on Watch sat on the hatch coaming, the Petty Officer of the Watch near him, with his feet hanging through the hatch through which the air for the gas engines was being drawn. I still wonder why I was not afflicted with rheumatism in spite of leather trousers. The third man on watch, a seaman, stood on a small three-cornered platform above the conning tower; he was lashed to his station in heavy seas.'

It may be seen therefore that a contemporary description of the U-boats as being 'horribly cramped, crowded and complicated internally' is highly accurate. They were also extremely

USN personnel in the torpedo compartment of a captured U-boat. *(via navsource.org)*

smelly, a combination of the body odour of unwashed men, their breath, rotting food, diesel oil, cooking odours, urine and excrement in the bilges. From U-19 onwards the Körting engines, which often produced clouds of undesirable exhaust smoke and occasional showers of sparks, were replaced by much more suitable diesel engines. They were also the first to have sufficient range to reach the Irish Sea or the Atlantic coast, being able to cruise on the surface 4000 nautical miles at her full speed of 15.4 knots, 5300 nautical miles at 9.5 knots or 7600 nautical miles at 8 knots. Underwater, U-19's maximum speed was 9.6 knots but this would drain the batteries within an hour. At 5 knots she could travel for 80 miles before having to recharge. Her diving time to periscope depth was 75 seconds which was reduced in successive U-boat classes to 30 or 40 seconds. Other improvements would include 'shark's head' bows surmounted by net-cutters and the provision of 'jumping wires.' A jumping wire was a wire cable stretched between the bow and stern of a U-boat, via the conning tower or periscope standards. Its purpose was to permit it to pass under nets, without it snagging on its superstructure, the wire causing the net to ride up and over the top of the submarine. Larger calibre guns were mounted, eventually from 3.5-inch (88 mm) and 4.1-inch (105 mm) up to 5.9-inch (150 mm), to save on the expenditure of torpedoes and allow combat on a more equal basis with surface ships.

German designers and shipbuilders were painstaking, well-organised and listened to feedback from the submariners. Two new types were introduced during the war, the UB (coastal) and UC (small minelayer) series, both of which were comparatively simple and suitable for rapid production. The mines would, in the first instance, sink to the seabed. A release mechanism would then activate which would cause the mine to rise to the surface, tethered to a sinker weight by a wire-rope mooring line, where it would unobtrusively await its prey. Spiess also wrote a brief and rather lyrical account of what it was like to submerge:

> 'It is an exciting moment when one stands for the first time in the conning tower and notes through the thick glass of the small ports how the deck becomes gradually covered with water and the bow slowly sinks. Have all the openings been properly closed and is the pressure hull tight? In the clear sea water, when the sun is shining, the silvery air bubbles sparkle all over the boat's hull and rise as in an aquarium. I was always interested in observing the surface of the water from below after diving as it looked like moving glass. Under favourable lighting conditions it is possible to see about twenty metres under water, but we were never able to see the bow of the boat. At times, when the boat was lying still on the bottom, we could observe fish swimming close by the ports of the conning tower, attracted by the electric light which was shining through. To put a submarine on the bottom without undue shock requires a certain amount of expert handling on the part of the Commanding Officer. The boat is brought down by the horizontal rudders until contact with the ground is slowly made; and then, after stopping the motors, the regulating [i.e. compensating] tanks are flooded until the boat is heavier by several tons and remains anchored of its own weight – that is, provided there is no heavy seaway or strong tide.'

One of Walther Schweiger's officers serving in U-20, Leutnant zur See Rudolph Zentner later described another aspect of U-boat life – the celebration of Christmas:

U-10 at full speed.
(Library of Congress)

U-14 *(Library of Congress)*

'The U-20 took a comfortable dive and settled on the bottom. And now, cried Commander Schwieger, "We can celebrate Christmas." The boat found a snug resting place on the muddy floor of the North Sea, and we were comfortably tied for the night sixty feet below the surface of the water. The tiny messroom was decorated in style. A green wreath hung at one end as a Christmas tree. We didn't have any lighted candles on it. They would have been too risky in the oil-reeking interior of a submarine. The tables were loaded with food. It all came out of cans, but we didn't mind that. That one night officers and men had their mess together. It was rather close quarters. We had a crew of four officers and thirty-two men. We were all in our leather submarine suits. It was no dress affair. No stiff bosoms, no tail coats. No "fish and soup" as you call them. In short, there were many drawbacks, but good spirits were not one of them. In the tight, overcrowded little mess room we ate and talked. The dinner was washed down with tea mixed with rum, and I lost count of the number of toasts that were drunk. After dinner came a concert. Yes, we had an orchestra. It consisted of three pieces, a violin, a mandolin and the inevitable nautical accordion.'

That conditions on board British submarines were equally spartan may be gauged from remarks made by Field Marshal Sir William Robertson, Chief of the Imperial General Staff. He asked the young commander of the boat he was inspecting if he liked the life. On being informed that he did, the Field Marshal replied with 'a grunt and a glance', 'Umph, well you're damned easily pleased.'

Chapter 3

The Development of Aerial and other U-boat Countermeasures

Airships

TO RETURN TO NAVAL aeronautical matters, the Admiralty was not, however, unmindful of the potential threat posed by a greater submarine force. The professional head of the Royal Navy, the First Sea Lord, Admiral of the Fleet Lord Fisher called a meeting at the Admiralty on 28th February 1915.

He requested proposals to enhance the capability of the Navy to provide surveillance and deterrence from the air. He had in mind a small airship type to the following specification:

(a) The airship was to search for submarines in enclosed or relatively enclosed waters.

(b) She was to be capable of remaining up in all ordinary weather, and should therefore have an air-speed of 40 to 50 mph.

(c) She should have an endurance at full speed of about eight hours, carrying a crew of two.

(d) She should carry a wireless telegraphy outfit with a range of 30 to 40 miles.

(e) She should take up about 160 lbs weight of explosives in the form of bombs.

(f) She should normally fly at about 750 feet altitude, but be capable of flying up to 5000 feet.

(g) The design was to be as simple as possible, in order that large numbers of these ships should be produced without undue delay.

Right: Admiral of the Fleet Lord Fisher. *(Library of Congress)*

Middle: Lieutenant Commander Neville Usborne. *(Usborne Family Collection)*

Far right: Thomas Cave-Browne-Cave. *(National Portrait Gallery)*

(h) An ample allowance of lift to be made for gas deterioration, so that each ship should remain in commission on one charge of gas as long as possible.

The new craft which had been designated by Lord Fisher as submarine-searching or perhaps submarine-scout would become known as the SS Class and would be crewed by the Royal Naval Air Service (RNAS). Considerable design input was made by two serving RNAS officers, Wing Commander Neville Usborne RN (who was an Anglo-Irishman born in Queenstown) and Flight Lieutenant Thomas Cave-Browne-Cave RNAS.

The prototype SS Class airship took to the air within a month. The envelope of a small non-rigid airship, which had been in storage at Farnborough, was married with the fuselage of a BE2c aircraft. A non-rigid airship has no keel or metal framework, as was the case with the much larger Zeppelins. This is also known as a blimp.

A fabric envelope or gasbag filled with hydrogen is kept firm and in streamlined shape by internal air-filled bags or ballonets. As hydrogen is lighter than air it can carry the weight

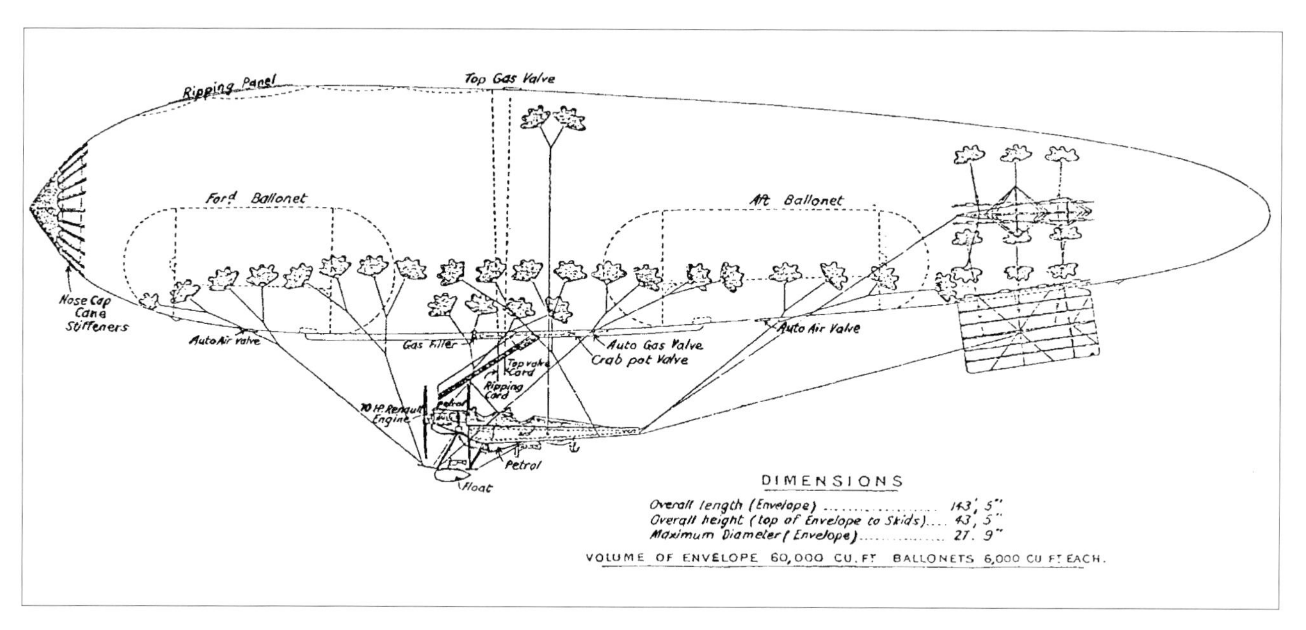

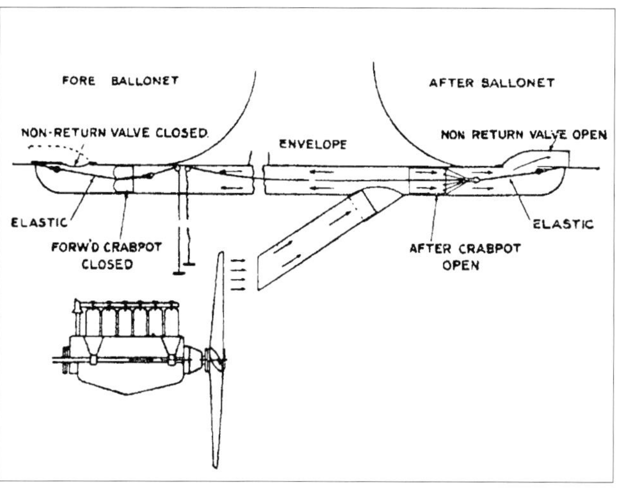

Above: The SS non-rigid blimps had two ballonets placed in tandem. The air was collected from the slipstream by a metal scoop and directed as required through manually operated valves into the ballonets, to maintain the internal pressure and so keep the envelope streamlined. Air could also be admitted selectively to control the trim; favouring one ballonet caused the airship to tilt.

Left: The utilisation of the propellor's slipstream to inflate the ballonets was a brilliantly simple solution to one of the main problems of non-rigid airship design. The method was devised for the SS prototype in 1915 and has since been used by nearly all British blimps up to the present day.

(Both diagrams by permission of Patrick Abbott)

of the whole structure and its contents – the rule of thumb was that 1000 cubic feet of hydrogen would lift 65 lbs in weight. Beneath the gasbag is suspended a car which carried the crew, engine, fuel and weapons. The purpose of the engine is simply to push or pull the airship through the air, lift is not provided by forward movement generated by the engine but by the gas inside.

The SS Class was a fairly simple and basic design but it had several merits. The production model met the specification as regards speed, 40–50 mph, and endurance, 8–12 hours. It could climb with its crew of two to a height of more than 5000 feet. It was also cheap, with a unit cost of £2500. The shape was reasonably streamlined, being blunt at the nose but tapering towards the tail. The gasbag had a capacity of 60,000 cubic feet and was 143 feet six inches in length, with a maximum diameter of 27 feet nine inches. As a comparison, a typical German Zeppelin airship of the period measured more than 700 feet in length, with a diameter of 78 feet and a gasbag capacity of 2,500,000 cubic feet. The two internal ballonets mounted fore and aft not only ensured that the envelope kept its shape but could also be used by the pilot to adjust the airship's trim. They were inflated by means of a metal scoop mounted to catch the slipstream of the propeller. Two non-return valves made from fabric, the 'crab-pots', controlled the flow of air into and out of the ballonets. The gross weight which the airship could lift, including its own structure, was 4180 lb, which gave a net lift available for crew, fuel, ballast and armament of 1434 lb. The disposable lift with a crew of two on board and full fuel tanks was 659 lb. The envelope was made of rubber-proofed fabric, reminiscent of an old-fashioned mackintosh. It consisted of four layers, two of fabric with a layer of rubber in between and on the inner surface. At the tail of the gasbag a single vertical fin and rudder were fitted ventrally, while horizontal fins and elevators were affixed to port and starboard.

Flight Commander Colmore (on right) watching final preparations. FSL York-Moore in the front cockpit is operating the 'crab-pot' valve rope. *(Yorke-Moore Family)*

To make them completely gastight and protected from the ravages of weather, salt water and sun, four coats of dope were applied to the outer surface, with a top coat being of aluminium varnish. The BE2c fuselage was retained stripped of its wings, rudders, elevators and eventually wheels, axles and suspension. Propulsion was by means of a 75 hp air-cooled Renault engine driving a large four-bladed propeller. The observer, who also operated the wireless set, sat in the front seat with the pilot behind him. It was powered by two four-volt accumulator batteries rather than by fitting a generator driven by the engine. This had two advantages, it was lighter and would still operate in the event of an engine failure. Communication was by means of Morse Code. The wireless telegraphy receiver and Type 52 transmitter had a range of between 50 and 60 miles, when flying at not less than 800 feet. A long trailing aerial some 200 feet in length with a lead weight on the end to keep it from fouling any part of the airship was wound down from a reel fitted to the side of the car. Small bombs, eight 16 lb or two 65 lb and a Lewis machine-gun could be carried by way of armament. A lever bomb-sight was fitted and the release was operated by Bowden wire control. It was considered to be, 'capable of being flown by a young midshipman with small-boat training.' To this end junior officers were brought in from the Grand Fleet by means of ships' Captains being asked to select one of his midshipmen who was willing to volunteer for 'special temporary and hazardous service.' One of these was Thomas Elmhirst, from the battlecruiser HMS *Indomitable*, whom we shall meet shortly. Enthusiastic, young, direct entry civilians were also induced to join up – specifically for this purpose. An intensive training course, lasting about a month, was given in the theory of aerostatics (aeronautics, navigation, meteorology, engineering, rigging and engine overhaul), practical balloon flying and mastering the controls of a small airship. Captain TB Williams, who trained at Wormwood Scrubs, recalled the, 'excellent instruction in knots, splicing and rigging, and I can still tie bowlines, sheet-bends and reef knots as a habit. A Warrant Officer took us into the mysteries of eta-patches, valves, sleeves, petticoats, crabpots, fabric and ballonets.' The young officers greatly enjoyed the balloon flying, eight qualifying flights – six under instruction (including a night flight), one trip as second in command and finally, a solo. A static balloon was available at Wormwood Scrubs for initial experience let up on a winch, 'sitting on a short length of planking like a bosun's chair with their legs dangling.' Free ballooning was undertaken from Hurlingham Polo Ground in the charmingly named, *Swallow*, *Shrimp*, *Salmon*, *Seahorse*, *Plover* or *North Star*, to name some of those available. Elmhirst's solo was to Suffolk when the only difficulty encountered was, after landing in a thorn hedge, when an inquisitive local with a smoking pipe in his mouth began to closely examine the gas valve aperture. He was dissuaded from this dangerous practice very quickly. The initial airship flying instruction was carried out in SS36, which served at Wormwood Scrubs from February 1916. One of the ballooning instructors at Wormwood Scrubs was Squadron Commander John Dunville, where his wife, Violet, also served, helping to run the canteen for the young officers. Dunville was from Holywood in Co Down, the son of a well-known and prosperous whiskey distilling family, which owned the Royal Irish Distillery in Belfast.

John Dunville in the basket of *Banshee*, probably taken under the Battersea Railway Arch, the Short Brothers' Balloon Factory. *(via Michael Clarke)*

He became a prominent member of the Aero Clubs of Ireland and the United Kingdom, taking delivery of his own balloon, manufactured at Eustace and Oswald Short's balloon factory at Battersea, in 1907. On 9th December 1907 he was awarded the ninth Aeronaut's Licence to be issued by the Aero Club of the United Kingdom. In February 1910 he had flown across the

Irish Sea by balloon from Dublin to Macclesfield. This was only the third crossing from Ireland to Great Britain by balloon. The young midshipmen profited not only from his skills but also from his 'ideal blend of kindness and strictness.' Tom Elmhirst states that:

> 'He was probably the most competent balloon pilot in England [sic] in the years before 1914. To us young officers he was a middle-aged captain. He amused us by wearing galoshes (he did not wish to get his feet wet) and carried a small bag containing a railway guide, so that he could at landing catch the nearest quick train back to London. He also, and we envied him, had a large, white Rolls-Royce, in which a chauffeur in daylight would attempt to follow his balloon flight.'

Admiral Fisher demanded that forty more of these small airships should be produced as expeditiously as possible. Neither the First Sea Lord nor the First Lord were to remain at the Admiralty long enough to see the SS Class into operation. In May, following the tragedy of the Dardanelles Campaign, Winston Churchill resigned and Jackie Fisher also departed. They were succeeded by the former Prime Minister, Arthur Balfour and Admiral Sir Henry Jackson respectively.

Luce Bay

Aerial surveillance would also play a part in the naval activities controlled from Larne. In the spring of 1915 an airship base was constructed on a 444-acre site at Luce Bay in Wigtownshire, on the coast of Dumfries and Galloway, four and a half miles south-east of Stranraer. It was organised on naval lines, with a small parade ground carefully surrounded by whitewashed stones and referred to as the 'Quarterdeck', a ship's bell and a flag-staff flying

An aerial photograph of RNAS Luce Bay, note the airship to the right of the picture and the pair of windbreaks at each end of the airship shed. *(JM Bruce/GS Leslie Collection)*

The view from an airship approaching Loch Ryan and the familiar landmark of the lighthouse on Corsewall Point. *(Donnie Nelson Collection)*

the White Ensign. At some distance from the accommodation and offices the land fell away in a gentle slope to the landing ground on which stood the airship shed, the gas-plant, tanks and power house. The shed was a lofty and substantial structure constructed of corrugated iron, measuring 300 feet in length, 70 feet wide and 50 feet in height, lit rather dimly by a row of windows in the upper part of the walls. It cost £8000 to build, with an extra £4200 for the windbreaks at either end. It was guarded at all times by a sentry whose task was to ensure that no smoking materials were taken inside. These were deposited in a cupboard to be reclaimed later. This was a very sensible precaution owing to the proximity of large quantities of hydrogen. It could hold up to four of the new SS Class. Beyond the shed lay a stretch of marshland and sand dunes around the bay.

While the facilities at Luce Bay were being built, Squadron Commander JN Fletcher was appointed Officer-in-Charge and during early March made a number of patrols over the North Channel from RNAS Walney Island at Barrow-in-Furness. He flew in the pre-war airship HMA No 17 *Beta II*. Fletcher was then appointed CO of Walney Island, Squadron Commander HL Woodcock having been posted to RNAS Kingsnorth in Kent. The first mention of the airship station in official correspondence from Larne was a note by Captain Gwatkin-Williams on 8th June, stating that the crews of the destroyers *Garry* and *Thorn* were presently having to do practically all the work when coaling at Stranraer, owing to the local labour being employed on the construction of 'an air base.'

At the beginning of August the first SS Class airships were brought to the little station at Dunragit by rail and then seven miles by road through winding lanes flanked by tall hedges or grassy banks, for inflation on site at Luce Bay. They were SS17 and SS23. The log books for both these airships have been preserved and on the inside front cover of SS17 the following pencilled notes are written, in the hand of Flight Sub-Lieutenant TW Elmhirst:

Beta II at Aldershot c1912. *(Author's Collection)*

Beta II in September 1912 with Lieutenant JN Fletcher RFC and Lieutenant Neville Usborne RN in the car. *(Author's Collection)*

- Care and Maintenance of Airships – Examination before flight
- Main and aux engines to be given a short run
- Amount of oil and petrol on board to be verified
- Equipment of spare parts, tools, fire extinguishers etc to be seen correct
- Rudders to be worked
- Ballast tanks filled
- All valves except top valves tested
- Trail and anchor rope to be seen ready for letting
- Propellers to be examined
- Instruments to be set
- Lighting plant to be tried
- Maps and charts to be seen on board

On 5th August SS17 was re-inflated at Luce Bay, having been flown for the first time at Walney Island on 3rd July, crewed by Lieutenant Ritchie and FSL Elmhirst. The next arrival was on 18th August when Midshipman EKH Turnour flew SS20 from Barrow at 4.50 am, arriving at Luce Bay at 7.30 am. The airships were secured in the shed by means of wire mooring cables, allowing the gas-filled envelopes to sway gently in the draught created when the sliding main doors were opened. The first operational flight from Luce Bay took place on 23rd August. It was a ninety minute patrol over the North Channel in SS17 made by FSL Tom Elmhirst. The log book noted, 'About 5.30 pm a message came through that there was a submarine in Luce Bay, SS17 was sent up as soon as possible and patrolled until dusk but sighted nothing.'

SS20 on test at Walney Island – note the third cockpit. *(via Peter Connon)*

In September, experiments were made concerning the carrying capacity of the airships and several flights were made with a pilot and two passengers. Numerous flights were made testing the airships' ability to carry and to drop bombs, as well as transmitting and receiving wireless messages. Turnour test-dropped a 20lb live Hales bomb over the sea on 16th September and on return from patrol the next day in SS20 bombed a 15 feet by 6 feet target from 500 feet, all 12 bombs falling within 20 feet of the target. Tom Elmhirst had added to his airship's equipment a bomb sight, which he had made himself – two nails in a piece of wood as foresight and backsight, the foresight movable to a scale marked with the speed of approach, 20 to 50 kts. 25th September was a significant day as, for the first time, all three airships, SS17, SS20 and SS23, were in the air at the same time. A young pilot, FSL TP Yorke-Moore, later noted, 'I was shown what to do by another pilot, himself a novice, and was asked if I thought I would be OK. I said I thought so and that was that.' The two patrols were made over the North Channel and the Irish Sea on 28th September and 1st October, each lasting nearly two hours. The potential of the airships was recognised as can be seen from the following extract. On 20th September Rear Admiral Boyle had written to the Secretary of the Admiralty from Larne Naval Base, 'I am requesting the Commanding Officer of the Luce Bay airships to confer with me as to the details of operations etc by these aircraft.' He added in a further report a week later, 'The three airships attached to the Luce Bay station have carried out several flights, and it is hoped that when the means of communication with these craft during flights has been improved, they may prove of use for scouting purposes. I am forwarding in a separate submission a request to have the W/T station now at Ballycastle, transferred to Larne, to enable me to communicate.' Activities in

Piloted by Flt Sub-Lt TW Elmhirst, SS17 takes to the air at Barrow on 3 July 1915. It was the first Vickers-built airship to fly and was sent to RNAS Luce Bay by rail. *(Ces Mowthorpe via JM Bruce/GS Leslie Collection)*

October continued on much the same basis, with one patrol being made as far as the Isle of Man and another over the Irish Sea with a duration of three and a half hours.

Larne

In the meantime on 12th September *Garry* and *Thorn* had escorted HMT (Hired Military Transport) *Olympic* from Belfast to Liverpool 'at a speed of 18-19 knots'.

The great White Star liner had just finished conversion to her new role as an armed transport for up to 6000 troops. Two weeks later she left Liverpool with a full load bound for the Gallipoli Campaign. *Tara* had been sent to Holyhead for a much-needed refit on 13th September, causing Rear Admiral Boyle a great deal of concern with regard to the diminution in number of available patrol vessels. In what would prove to be a valedictory report Gwatkin-Williams summarised *Tara's* effort and also returned to a previously expressed subject:

> 'Volume of shipping in North Channel increasing daily. *Tara* has been four months at sea, with all boilers alight since last refitted. I do not consider it prudent for her to continue longer without boilers being examined and cleaned and machinery opened up and refitted. During that time she has steamed 25,867 miles, and 59,671 miles since 18th August 1914, when she first put to sea at beginning of war. During all this period she has only had one small breakdown to steering telemotor. Some efficient sea keeping vessel should assist *Tara* in North Channel patrol duties. The two

destroyers are useful in fine weather, but are constantly absent on escort duties and making good defects. Trawlers are practically useless, owing to the lack of experience of many of the Skippers, their slow speed and their inability to signal or communicate rapidly. No confidence can at present be placed on North Channel patrol detecting in all cases vessels slipping through during the dark hours or hazy weather nor can the increasing volume of shipping be adequately overlooked.'

RMS *Olympic* *(Library of Congress)*

Tara did not return to Larne, for on receiving orders to sail for the Mediterranean, she was sunk off the coast of Egypt by a torpedo from U-35, Korvettenkapitän Waldemar Kophamel, on 5th November. Gwatkin-Williams would return to Ireland in March 1919, to give a very well-received lecture in Belfast on 'The *Tara* Adventures in the Mediterranean.'

Admiral Boyle supported Gwatkin-Williams, writing to the Secretary of the Admiralty:

> '*Tara* left North Channel Patrol for refit at Holyhead on 13th. She has since been ordered to proceed on foreign service on completion of refit. I have represented to Their Lordships, in submission No 537/8/1 of 16th instant, the necessity of having a sea keeping vessel allotted to the North Channel Patrol for the winter months, when the destroyers are likely to require to shelter frequently. The strengthening of the North Channel Patrol to warn shipping off the prohibited area seems necessary, in view of the danger to valuable ships and cargoes, if they continue to trespass and become involved in the nets as in the past, when the nets are fitted with the EC mines.'

An Indicator Net Type Mine was a metal cylindrical mine, with a conical top, around which were the electrical contact points, and with a conical base. The electrically fired device was invented by Admiral of the Fleet Sir Arthur Wilson VC. A moored mine only guarded a space equal to its own dimensions, since it exploded only on contact. Net-mines were hung in a net in such a way that when a submarine fouled the net, the drag exploded the attached mine. The effective area of the mine was therefore extended to that occupied by the net, for when a net enveloped a U-boat, the mine was brought with more or less certainty into contact with her hull. Boyle continued:

> 'The parting of all 9-inch hawsers at the salvage work on *Clementina* in Red Bay,

and the impossibility of obtaining further hawsers of this size, has caused the abandonment of the efforts to lift her farther, and steps are being taken to make the vessel tight, after getting her upright, as soon as the present easterly weather permits. The tugs and lumps being no longer required, were sent to Liverpool on 19th instant. The suggestion of the Commander of *Garry* as to the provision of more lights on the Irish Coast is being gone into, and will be put forward separately after consideration.'

Garry's Commander Werden Wilson had a suggestion of his own to make:

'The submarine reports furnished from the intelligence centres are not sent to the Air Station at Luce Bay. The destroyer standing off is also not informed, with the result that on going to sea the commanding officer often passes through an area where a submarine or a supposed submarine has been reported recently while he himself is in complete ignorance of the fact. It is suggested that the Air Station be furnished with all these reports and that they be directed to inform the destroyers in Stranraer by telephone to number 38 Stranraer. That is the number of Stranraer Harbour.'

Considerable progress was being made with improving the efficiency of working with the drifters in harbour, as this report to the Rear Admiral illustrates:

'It is not known how many Drifters are intended to be fitted with Electric Contact Mines. It is submitted that until experience in the use of these mines has been gained at this base, that 30 Drifters with nets so fitted is the maximum desirable. The Quay used for Naval purposes here gives Berths with a total length of 300 feet. This is used by all Auxiliary Patrol vessels based at Lame for coaling, ammunition and storage, and also affords a landing for the Harbour Boat Service. Without seriously hampering general efficiency, it is found impossible to fit out vessels with Nets etc at the Quay. The method adopted for fitting out and maintaining a supply of Indicator Nets for the Drifters is as follows. All nets are fitted, entirely by Drifters' Crews, at a field which situated at a distance of 800

Making an anti-submarine dectector net. *(Author's Collection)*

> yards from the Quay. A supply of fitted nets is taken to the Quay daily by hand cart. The Net Supply Drifter goes alongside the Quay daily, before loading operations are commenced, and takes in all fitted nets, at the same time landing damaged nets or nets for overhaul. This Drifter is anchored in a convenient berth, where vessels requiring new nets go alongside; during the remainder of the day she goes to vessels in the Repair Berth, under periodical overhaul, and exchanges their old nets for new or refitted ones. This system concentrates the work rendering strict supervision possible, and has answered remarkably well. Two Drifters are being specially fitted with stowage for nets, EC Mines and Circuits, and it is proposed to use the same system, with necessary modifications, for the supply and upkeep of EC Mines and Nets. It is hoped that a maximum of four may be worked simultaneously. It is thought that by careful organization, good fitting and workmanship, in the first place, there will be little difficulty in maintaining 30 fleets of Nets, so fitted, efficiently in use. The Field, which has a partly ruined Coast Guard Station and RNR Drill Battery in it, has proved a boon to the Drifters. In addition to the work of Indicator Nets and Buoys and EC Mines, it is used for instructional purposes, also for sailmaking and rigging. With the rough means at hand, progress has been made in making the Drifters self maintaining, excepting Shipwright and Engine Repairs. There are now sufficient men at this base to fit and work 30 Drifters with EC Mines and Nets.'

At the end of the month the salvage effort on *Clementina* was formally abandoned. Some indication of the scope of the task at Larne may be gained by examining the log of a typical day, 26th September, when 73 ships passed through the North Channel. In October the Weekly Report to the Admiralty from Larne noted the work being carried out by three ex-civilian motorboats. *Papakura*, Lieutenant Walter Brook, was stationed at Donaghadee, patrolling up to six hours a day. *Olivia*, Lieutenant A Paclock, remained in Belfast Lough, boarding vessels which had been 'sent to Belfast' by North Channel patrol vessels for having broken regulations as regards paperwork or had strayed into restricted areas, for example being within four nautical miles of Rathlin 50 minutes after sunset. The third motorboat, *Ian*, Sub Lieutenant Thomas A Walker, was based at Portaferry at the mouth of Strangford Lough, examining all shipping taking passage through the lough. On 15th October the patrol force out of Larne was augmented by the additional six Armed Trawlers of Unit 95, *Earl Lennox*, *Neptunian*, *Prince Victor*, *Strathmartin*, *Vesper* and *War Lord*.

Bentra

Consideration had been given to what Luce Bay based airships would do in the event that unfavourable weather prevented returning to base. Indeed as early as 20th July 1915 a minute from the Director Air Department at the Admiralty, Commodore Murray Sueter, had stated, 'With reference to the approval to establish airship patrols in the North Channel and Irish Sea ...a small portable shed has been approved at Larne for emergency use of these ships and this will be erected shortly.'

An airship lands at Bentra, Whitehead. *(D&N Calwell Collection)*

A mooring out station was established at Whitehead, at the head of Larne Lough, some eight miles south of Larne. This may be dated from the fact that a bank account was opened in the name of the RNAS at the Northern Bank in Whitehead on 14th October 1915, under the authorised signature of the Officer Commanding, Sub-Lieutenant Archibald Creighton. The land belonged to a local farmer, James Long of Bentra, a mile and a quarter out of the little seaside town and popular resort, which had a population of about 1500.

Queenstown and naval aviation

Vice Admiral Bayly had welcomed the arrival of the airships and may well have discussed their potential when he visited Kingstown, Larne and Buncrana in HMS *Adventure* in the autumn of 1915. He had another aviation asset under his command between August and December, the seaplane carrier HMS *Empress*, which was based at Queenstown on an experimental basis.

Short 830 Type Seaplane. *(Bombardier Belfast)*

The former South East and Chatham Railway cross-channel packet was at first fitted with canvas hangars and took part in the Cuxhaven Raid on Christmas Day 1914. A few months later, she was taken in hand at Liverpool and a permanent hangar for four aircraft replaced the canvas structures. Once at Queenstown, Bayly remarked favourably upon her four 135 hp Short Type 830 seaplanes carrying out bomb-dropping exercises. In a letter to the Admiralty dated 30th December, he stated that the often

HMS *Empress (Author's Collection)*

rough seas off the Irish coast would inhibit successful seaplane operations and presciently recommended the development of flying boat bases at Queenstown and Berehaven. *Empress* left Queenstown for the Mediterranean in December.

The first crossing to Ireland by airship

An incident of note took place on 3rd November, when Elmhirst, accompanied by W/T Operator Hall, had a very unpleasant experience in SS17. They ascended from Luce Bay at 10.27 am to carry out a patrol. The rudder controls of the airship failed over the North Channel, and trailed below the tail, out of reach from the car, rendering the airship as subject to the whims of the wind as a balloon. The wind was blowing easterly and off-shore, conditions were very bumpy. As the airship drifted towards the Irish coast, Elmhirst alighted on the water close to a trawler, whose skipper he asked for help. The crew simply ignored his entreaties. Even worse the propeller blades had touched the sea and had broken, forcing Elmhirst to switch off the engine. Ascending again, SS17 was blown towards another fishing vessel. Descending once more no helpful response ensued. With all its water ballast gone the airship was in a perilous position, so Elmhirst and his observer threw anything removable overboard, including the bombs, radio set and compass – to gain as much height as possible. A 40lb bag of sand ballast, the grapnel and its long trailing rope were the last to go over the side. By this means it was possible to climb over the 300 feet high sheer cliffs to the north of Larne and to make a landing at 4.15 pm

An official note concerning Tom Elmhirst's flight. *(National Archive)*

FILE

① SS Airship patrolling from LUCE BAY over Northern Channel. Rudder controls carried away over channel and trailed below tail out of reach of car. Ship being out of control pilot stopped engine and ballooned. Wind N.E. about 10 m.p.h. Pilot landed alongside trawler & signalled for a boat to come & recover controls. Trawler failed to acknowledge signals although well within earshot & saw signals plainly. Pilot discharged all ballast, left water and realighted near another trawler. Same programme same result. Discharged W/T gear, bombs etc and just landed in Ireland. On a report being called for from trawlers as to why no assistance was rendered the following was received

"We did observe a British Airship manoeuvring. It was a very pretty sight"

in Laing's Field off the Ballytober Road, by pulling on the ripcord to totally deflate the envelope, after a flight lasting over five hours. A local farmer and his wife, by the name of Smiley, helped the aviators to un-rig and pack up the downed craft and then provided very welcome refreshments. The Senior Naval Officer sent a couple of lorries to collect them but gave them rather a grumpy greeting in comparison to the farmer and his wife. Two Armed Trawlers, *Earl Lennox* and *Neptunian*, returned the envelope, rigging and car to Luce Bay on 4th November.

The fishing trawler captains were subsequently asked why they had been so unhelpful. The reply came, 'We did observe a British airship manoeuvring; it was a very pretty sight', which may give some clue as to their political sympathies. Be that as it may, Elmhirst was to assert later that he had made the first crossing from Great Britain to Ireland by airship – a balloon had first made the journey across the Irish Sea on 18th July 1817, by which means Windham Sadler was carried from Dublin to Holyhead. Then on 29th November 1907, the French airship, *La Patrie*, was blown unmanned from her moorings at Verdun across Northern France, Cornwall and the Irish Sea. Numerous sightings were reported in the *Belfast Telegraph*, 'to the consternation of the inhabitants' of country towns and villages who 'gathered in large numbers' to view the great yellow dirigible pass overhead.

The propeller lost by *La Patrie* in Co Down. *(Author's Collection)*

She struck a hillside on the south side of Belfast Lough at Ballydavey, Holywood, County Down, losing a propeller in the process but ascended once more and was last seen off the Isle of Islay speeding into oblivion.

On the same day as SS17's adventures, SS23 made a two-hour patrol during which 'three schooners were examined'. Progress was hindered by a spate of forced landings due to trouble with the temperamental engines. A great deal of experimental work was also being carried out with regard to the best way to rig the airships and their 'cars'.

The residents of Larne were able to have sight of a much larger airship on 3rd November 1915 when HMA No 4, under the command of Flight Lieutenant FLM Boothby, flew directly from Walney Island, near Barrow-in-Furness, to the County Antrim coast. This non-rigid airship, which had been ordered before the war by the Admiralty from the German Parseval company, had a cubic capacity six times greater than the SS Class. It carried a crew of nine in a roomy open car made from duralumin and was powered by two 170 hp Maybach engines.

During the autumn of 1915 it made several training and proving flights as far as north-east Antrim, the Mull of Galloway, the Isle of Man and Anglesey. It was noted in the log at Luce Bay on 12th November that, 'No 4 Naval Airship passed over north end of the Isle of Man about 1.30 pm and after going to Larne Harbour proceeded back to Barrow'. HMA No 4 had given useful operational service early in the war, escorting troopships carrying the British Expeditionary Force (BEF) to France. From 1915 until 1917 she was used for training.

HMA No 4 takes to the air. *(JM Bruce GS Leslie Collection)*

One of the unsung but very useful vessels based at Larne, arriving there on 13th November, was HMY *Zara*, Acting Commander Henry Smith RNR. She was an elegant armed yacht which was used for patrol duties and conveying orders and route warnings to passing traffic. She was armed with two 3-inch guns and depth charges, and would be equipped with hydrophones. Chatterton writes of the steam yachts, 'Perhaps of all the sea-going patrol craft these long, narrow pleasure ships, mostly single-screwed, with big turning circles, were the least satisfactory, simply for the reason that they had been built for enjoyment and not hard wear. All the same, their officers and crews obtained excellent results with them, and the wintry weather tried ships and men pretty thoroughly.'

In the middle of November the airship crews were engaged in painting red, white and blue identification stripes on the rudder, with blue next to hinge.

Then later in the month Tom Elmhirst and SS17 flew on one day seven altitude tests up to 3100 feet with 175 lbs ballast. Some diversion was found in the last week of November when several visiting Italian airship officers were taken for flights round the area. One of the officers was Capitano di Fregata Salvatore Denti Amari, Duca di Pirajno, soon to become Chief of Staff of the Italian Airship Section.

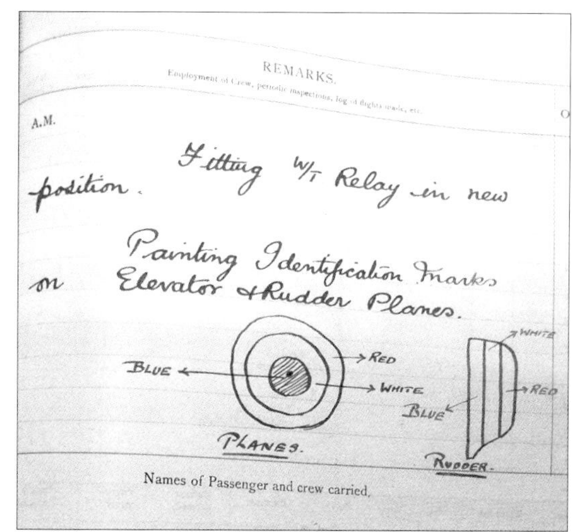

REMARKS.

A.M.

Fitting W/T Relay in new position.

Painting Identification Marks on Elevator & Rudder Planes.

BLUE
RED
WHITE
PLANES.
WHITE
RED
BLUE
RUDDER.

Names of Passenger and crew carried.

A page from the logbook of SS23 dated 16 November 1915. *(National Archive)*

Chapter 4

From the end of 1915 to the start of 1917

THE WINTER MONTHS OF 1915-16 brought very poor weather which greatly hindered the fledgling operation at Luce Bay and considerably reduced the possibility of making any patrols at all.

A Coastal Motor Boat similar to *Patricia*. *(Author's Collection)*

In the middle of November HM Motorboat No 8, *Patricia*, Lieutenant CLL Ionides, was based nearby at Drummore and was used mostly for target towing and local off-shore patrols, the station log recording, for example, '18th November 1915: *Patricia* put out to sea at 10.30 pm looking for a submarine which was reported in Luce Bay.' It was thought that the Germans might mount a sea-borne raid on the base, with troops brought to the Scottish coast by U-boats but they were never so bold or adventurous. *Patricia* would also be used for towing targets for Lewis gun and rifle practice. More warlike preparations took place on 17th December when, in the course of the usual bomb dropping practice, SS17 and Tom Elmhirst made, 'six flights, including two live bombs (Hales Mk I and II) on flights 5 & 6 from height of 500 feet, one being within 20 feet of target.' The final patrol of the year was on 28th December when FSL Yorke-Moore took SS17 out for a two hour sortie.

SS23, piloted by Flt Sub-Lt TP Yorke-Moore, approaches over Torrs Warren prior to landing at Luce Bay. *(Donnie Nelson Collection)*

By early 1916 the station had a naval complement of about 40, half of whom were officers. In February it was noted that a new envelope had arrived for SS23 and that one was awaited for SS33. On 20th February two night flights were carried out in SS17, the first by Elmhirst and Yorke-Moore and then Elmhirst and Turnour.

The configuration of the Maurice Farman type fuselage is clearly shown in this picture. *(Airship Heritage Trust)*

SS30A at Cranwell in 1918, a good close-up of the Maurice Farman car. *(Ces Mowthorpe Collection)*

It was not until the middle of March 1916 that any serious aerial activity resumed. This followed on from the arrival of two more airships, SS33 and SS38. These were different in that their cars resembled the fuselage of a Maurice Farman aeroplane and were fitted with pusher rather than tractor propellers – which had the advantage of enabling the crew to avoid buffeting from the slipstream.

They were slightly slower than the BE2c models but the cars were somewhat roomier and more comfortable for the crew. The pilot sat in the front seat, with the wireless telegraphy operator/observer behind. Dual flying controls were provided. A third seat could be installed for a passenger or engineer. Engine reliability was a problem with the SS Class, faulty cylinder heads being a particular aspect. At times running repairs were even carried out in flight, standing on the skids, holding on with one hand while working with the other. Unbelievable as it now sounds, the engine was then restarted by a hefty swing on the propeller. In March it was noted that 'high winds and unfavourable weather have made lengthy patrols impossible during the last 30 days', however eight hours of patrolling were carried out on 18th March and no less than 21 instructional flights on 17th, 'including live and

Below: A plan drawing of the SS Class airship with a Maurice Farman car. *(Author's Collection)*

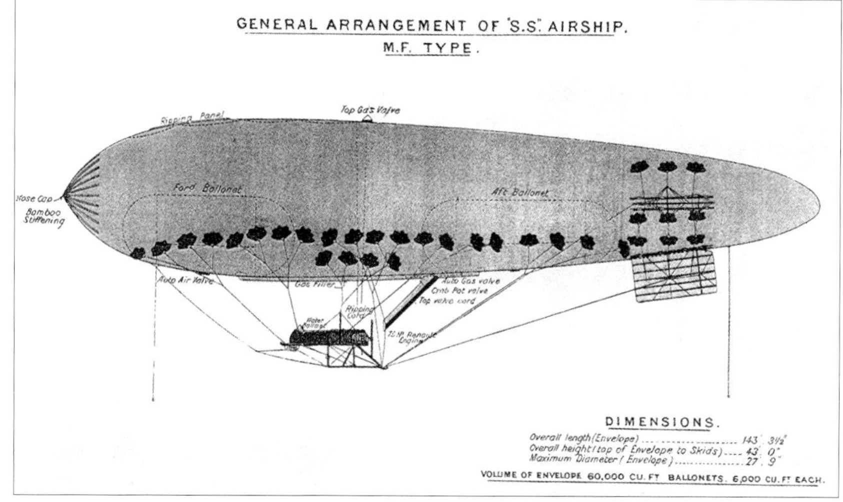

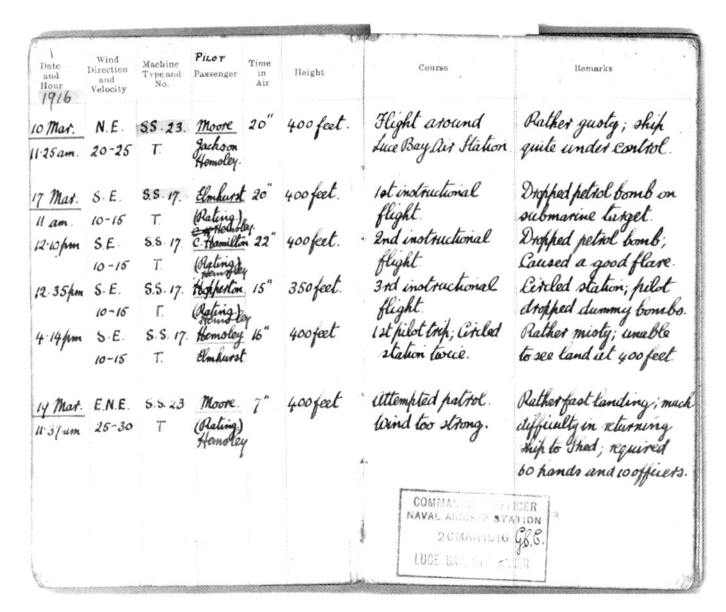

Date and Hour	Wind Direction and Velocity	Machine Type and No.	Pilot Passenger	Time in Air	Height	Course	Remarks
1916							
10 Mar. 11·25 am.	N.E. 20-25	SS.23. T.	Moore Jackson Hemsley	20"	400 feet.	Flight around Luce Bay Air Station	Rather gusty; ship quite under control.
17 Mar. 11 am.	S.E. 10-15	SS.17. T.	Elmhurst (Rating) Hemsley	20"	400 feet.	1st instructional flight	Dropped petrol bomb on submarine target.
12·10 pm	S.E. 10-15	SS.17 T.	C. Hamilton (Rating) Hemsley	22"	400 feet.	2nd instructional flight	Dropped petrol bomb; Caused a good flare.
12·35 pm	S.E. 10-15	SS.17. T.	Hopperton (Rating) Hemsley	15"	350 feet.	3rd instructional flight	Circled station; pilot dropped dummy bombs.
4·14 pm	S.E. 10-15	S.S.17. T.	Hemsley Elmhurst	16"	400 feet	1st pilot trip; Circled station twice.	Rather misty; unable to see land at 400 feet.
14 Mar. 11·31 am	E.N.E. 25-30	S.S.23 T	Moore (Rating) Hemsley	7"	400 feet	Attempted patrol. Wind too strong.	Rather fast landing; much difficulty in returning ship to shed; required 60 hands and 10 officers.

A page from Flight Sub Lieutenant Bernard Hemsley's Log Book. *(National Archive)*

dummy bomb practice – out of three live bombs dropped from 400 feet, two were bulls, the third being 14 feet short and to the right.' New, larger patrol motor launches were also based at Luce Bay, firstly No 248 and then in March, following engine troubles, No 546, which proved to be more reliable. Chatterton describes the role of motor launches as follows, 'Patrolling, chasing submarines, escorting shipping, exploring minefields, sweeping up and destroying the mines themselves – there was hardly a maritime job which they failed to tackle.'

The oldest airship, SS17, had been relegated to training duties since the end of 1915 and was deflated in April before being sent to Wormwood Scrubs Depot. It is of interest to note that activities were also curtailed by the fact that the crew of the *Princess Maud* were on strike for most of April, owing to a wages dispute. Four airships were now operational SS20, SS23, SS33 and SS38. It was reported on 22nd April that U-boat wireless messages had been intercepted at a distance of 100 miles and that the station was 'almost completed.' A patrol report on 26th April made by FSLs EK Turnour and BW Hemsley in SS20 noted two sightings, a large, three-masted sailing vessel 15 miles WNW of the Mull of Galloway and two tramp steamers further to the west.

Moreover, in April 1916, U-22, under the command of Kapitänleutnant Bruno Hoppe, sank 11,000 tons in the seas between Ireland and Great Britain and had a narrow escape

Form.—Air I.O. 1.

SECRET.

No. of Reconnaissance	Date	Airship No.	Type	Squadron	Pilot	Observer	Ref. Map	Hour at which reconnaissance commenced	Hour at which reconnaissance concluded
2	26/4/16	20	S.S.	Luce Bay	Sub Lieutenant E.K.Turnour, R.N.	Flight Sub Lieut. B.W.Hemsley, R.N.	X.124	5.45 p.m.	7.45 p.m.

Time	Place	Observations and weather conditions
6.20 p.m.	15 miles W.N.W. of Mull of Galloway.	Large three-masted sailing vessel.
6.48 p.m.	20 miles W of Mull of Galloway.	Two tramps. Visibility good. Weather:- B.C.M. Wind:- S.W. 10 to 15 M.P.H.

AIR DEPARTMENT. A.M. 28 APR 1916

Signed B.W. Hemsley Flight Sub Lieut RN. Observer.

A patrol report of April 1916 by Bernard Hemsley. *(National Archive)*

from being rammed by a cruiser off Belfast. Her commander had a hair-raising experience when the cruiser loomed out of the mist. Second Officer, Oberleutnant zur See Ernst Hashagen, was on watch and as the cruiser opened fire he ordered the submarine down in a crash dive.

The U-boat tilted violently as the diving planes angled down and the rush of water into the ballast tanks roared like thunder. Hashagen had given orders to level out at 50 feet. But due to an unexpected fault in the depth rudder, the they plunged below the 50 feet mark and kept falling, 'she tilted up and down like a rocking-horse, sinking now by the head and then by the stern – but always sinking.' Hoppe rushed into the control room but with the British cruiser lying in wait on the surface he did not dare to blow the ballast tanks and was at the mercy of her broken rudder. At 200 feet the pressure of the sea would crush the steel hull like a matchbox. Already the metal was groaning under the strain, the steel support beams were beginning to buckle and tiny beads of water glistened around unseated rivets. But something even more terrifying was happening in the battery compartment. 'Everything else lost its importance … I caught the acrid smell of chlorine gas and everyone was coughing, spluttering and choking,' Hashagen recalled. Sea water, forcing its way through the seams, had mixed with the sulphuric acid of the batteries and was sending off clouds of greenish-yellow vapour. Hashagen continued, 'I don't think there is anything that will strike such fear in a submarine man as the thought of being trapped in the iron hull while choking gas seeps from the batteries bit by bit. No death could be more agonizing.'

3 Date and Hour	Wind Direction and Velocity	Machine Type and No.	Pilot Passenger	Time in Air 2' 48"	Height	Course	Remarks
25 April 3·54 pm	S.S.W. 5-10	S.S. 23 T.	Hamilton Hemsley	11"	300 feet	Vicinity of Aerodrome	Dropped dummy bombs. Exceedingly misty.
26 April 5·45 pm	S.W. 10-15	S.S. 20 T.	Turnout Hemsley	2 hrs.	800 feet	W.S.W. to Mull of Galloway, E, over Scare Rocks.	Patrol over Luce Bay + Irish Channel.
29 April 9·54 am.	E.N.E. 7-15	S.S. 23 T.	Hamilton Hemsley Goodwin	16"	750 feet	Steering east; turned to west over Dunragit	Engine test. Very "dunty" Heavy landing.
3·5 pm.	Lt. Airs	S.S. 38. P.	Jackson Hemsley.	10"	300 feet	Around Aerodrome	Slightly heavy landing. Ship had little lift.
4-30 pm	Lt. Airs	S.S. 38. P.	Beaumont Hemsley	10"	400 feet.	Around Aerodrome	Heavy landing. No 1 [landing] starboard suspension broke on

COMMANDING OFFICER NAVAL AIRSHIP STATION 7 MAY 1916 LUCE BAY, STRANRAER

COMMANDING OFFICER NAVAL AIRSHIP STATION 30 APR 1916 LUCE BAY, STRANRAER

Don't be so critical

Another page from FSL Hemsley's Log Book. *(National Archive)*

Oberleutnant zur See Ernst Hashagen *(Author's Collection)*

Faced by the lethal vapour which was billowing and swirling through the boat Hoppe decided to stake all their lives on a desperate gamble. The ballast tanks were blown and U-22 swept to the surface like a cork. All thoughts of the British cruiser had gone in the face of the crying need for air. Hashagen again, 'Better to be shot to pieces and drown in a quiet way than this death by choking torment.' The fog was still thick when they broke surface and although only a few hundred yards away the enemy warship did not notice the U-boat rolling gently in the swell with her hatches thrown open as the life-giving fresh air was sucked down into her hull. There was a tense interval while U-22 cleared her compartments of gas and then, using the electric motors, Hoppe crept silently away into the mist leaving his adversary unaware that its prey had escaped from under its very nose.

SS23 takes to the air. *(Yorke-Moore Family)*

On 1st May at Luce Bay an experiment was made in dropping signal flares. A much more detailed report was completed by FSL John Cole-Hamilton, who had replaced Yorke-Moore, flying SS23 by himself on 4th May. He left the ground at 6.52 am, within half an hour he passed over a small steamer 'black hull and funnel, white upper works, red waterline, flying no flag' and 'a brigantine, white hull, brown sails, all sails set, flying no flag.' After an hour in the air, with Great Copeland Island three miles abeam, he descended to inspect, 'a white object' which turned out to be a small pleasure motor boat. Forty-five minutes later abeam of Muick (sic) Island (probably Muck Island, Islandmagee) he, 'passed over Swedish Tank steamer SS *Nyland*. Course approximately E by N, speed 6 knots.' As he made his way across the North Channel he saw a steam coasting vessel, 'black hull, yellow funnel, black top, flying Red Ensign'. He landed back at Luce Bay at 9.50 am. Turnour made a patrol of similar duration in SS20 on the same day, spotting three steamers, two schooners, a trawler, HM Minesweeper 963 and 'one of HM ships in tow, escorted by two trawlers.' His route took him by Portpatrick, Black Head, the Copeland Islands and Larne. Towards the end of May in visibilty which Cole-Hamilton recorded as 'Extraordinary' he flew over 'a White Star liner, two funnels; under escort of a destroyer, course south 15 knots.' A brief entry on 10th May noted, 'Longest patrol so far five hours FSL Cole-Hamilton with AM Sinclair.'

The post of Senior Naval Officer in Larne was assumed by Commodore Sir Alfred W Paget, RNR. He was another distinguished Admiral who had resumed service from the

The officers and men of RNAS Luce Bay in 1916. *(Elmhirst Family)*

retired list and was 64 years old in 1916, 'a fine English gentleman, and one of the bravest seamen who ever stepped on board a ship. With his monocle and white beard he was a picturesque personality.'

A hydrophone in use aboard a warship. *(Author's Collection)*

Other devices and weapons which would be deployed by vessels based in Larne from 1916 and onwards included the hydrophone and the depth charge. The hydrophone consisted of a receiver which, when lowered into the water, picked up the vibrations of a U-boat's propellers and passed the sound by means of 'communicating wires and headphones to a trained listener stationed on deck.' These were non-directional at first, until the development of directional hydrophones in 1917. An Ulsterman from Hillsborough, Lieutenant Hamilton Harty RNVR, renowned as the conducter of Manchester's Hallé Orchestra, used his musical ear to great effect tuning the equipment and matching pairs of hydrophones.

The depth charge (or wasserbombe as it was known to the U-boat crews, and ash can as referred to by the US Navy) was a 300 lb bomb fitted with a hydrostatic device to detonate it at a pre-set depth, therefore allowing a submarine to be attacked when submerged. The development of depth-charge throwers rather than simply rolling them over the stern increased their range and efficiency and chances of success. However, until well into 1917 the production rate was scandalously slow, with only 140 per week being manufactured as late as July of that year. By the end of the year output was up to 800 a week and climbing. During 1917 between 100 and 300 depth charges were expended per month but during the final six months of the war, on average, 2000 depth charges per month were dropped in attacks on U-boats. Johannes Spiess later wrote about the impact the initial use of depth charges made:

A Royal Navy destroyer dropping depth charges. *(Author's Collection)*

> 'The report was of a new danger we would have to surmount, a new and potent piece introduced onto the chessboard of war under the sea. Kapitänleutnant Richard Hartmann's U-49 had been bombed with depth charges. No such thing had been encountered before and the news made quite a sensation. And the more we thought of it, the less we liked it. The depth charge was a bomb loaded with two hundred pounds or so of high explosive. It could be set to explode at any desired depth under water when dropped overboard. At any place where a submarine was suspected an enemy ship dumped over its stern quantities of these infernal charges set to go off at

Above: The monitor HMS *Terror*. *(Author's Collection)*

Above right: RMS *Aquitania (Courtesy Clyde George Collection)*

various depths. A U-boat under water was peppered with a shower of them. If one exploded close enough it would sink the craft, or would at least make it leaky by springing the seams, and thus disable it. It was an evil invention and one destined to become part of our daily experience. The depth charge was aimed by sundry indicators that indicated the presence of a submarine, the sight of a periscope, bubbles sent up by the firing of a torpedo, the torpedo's track, the streak of oil on the surface left by leaking oil tanks, and so on.'

Patrols continued over the summer months but at a fairly low level. On the evening of 23rd June 1916 FSLs Hemsley, Elmhirst and GA Wearham made two short flights in SS33, firing ten rounds from a rifle at a ground target. In July Turnour flew over a, 'Norwegian steamer SS *Dana* carrying wood, black hull, two masts, yellow funnel, steering south' off the Copeland Islands and soon thereafter a 'Trinity House boat steering north.' He saw nothing further of note off Black Head or The Maidens but seven miles northwest of Corsewall

Form.—Air I.O. 1.

SECRET.

No. of Reconnaissance	Date	Airship No.	Airship Type	Squadron	Pilot	Observer	Ref. Map	Hour at which reconnaissance commenced	Hour at which reconnaissance concluded
17.	29/7/16.	20.	S.S.	Luce Bay.	Sub Lieutenant T.W. Elmhirst, R.N.	Sub Lieutenant T.W. Elmhirst, R.N.	X124.	10.5a.m.	2.55 p.m.

Time	Place	Observations and weather conditions
10.17 a.m.	Passed over Stoneykirk.	
10.32	Over Coast line 3 miles S. of Portpatrick.	
11.43	Over Copeland Lighthouse.	
12.35 p.m.	Over entrance to Belfast dock.	Observed H.M.S. "Terror" and S.S. "Aquitania".
12.53	Copeland Is. Lt. 2 miles on S. beam.	
1.15	5 miles S. 75 W. from Portpatrick.	Passed over large tramp steamer Juria steering N. by E.
1.20	Over Portpatrick Harbour A/c S. 85 W.	
1.40	A/c down centre of channel.	
2.22	Off Logan Bay.	Observed floating wreckage.
2.33	10 miles W. of Mull of Galloway.	Observed large oil tank steamer steering S.E., no name painted on. Black with black funnel and white upper works.
2.55	Landed.	
		Weather:- B.C. and O.C.
		Wind:- W. veering to S. up to 22 m.p.h.
		Visibility:- Fair.

Page No.

NAVAL AIRSHIP STATION
29 JUL 1916
LUCE BAY, STRANRAE

M

noted [illegible] 2.8.16.

Signed [illegible] for Pilot.

[3482] G179 10m 3/16 2869 G & S 111

Patrol Report by Tom Elmhirst in July 1916. *(National Archive)*

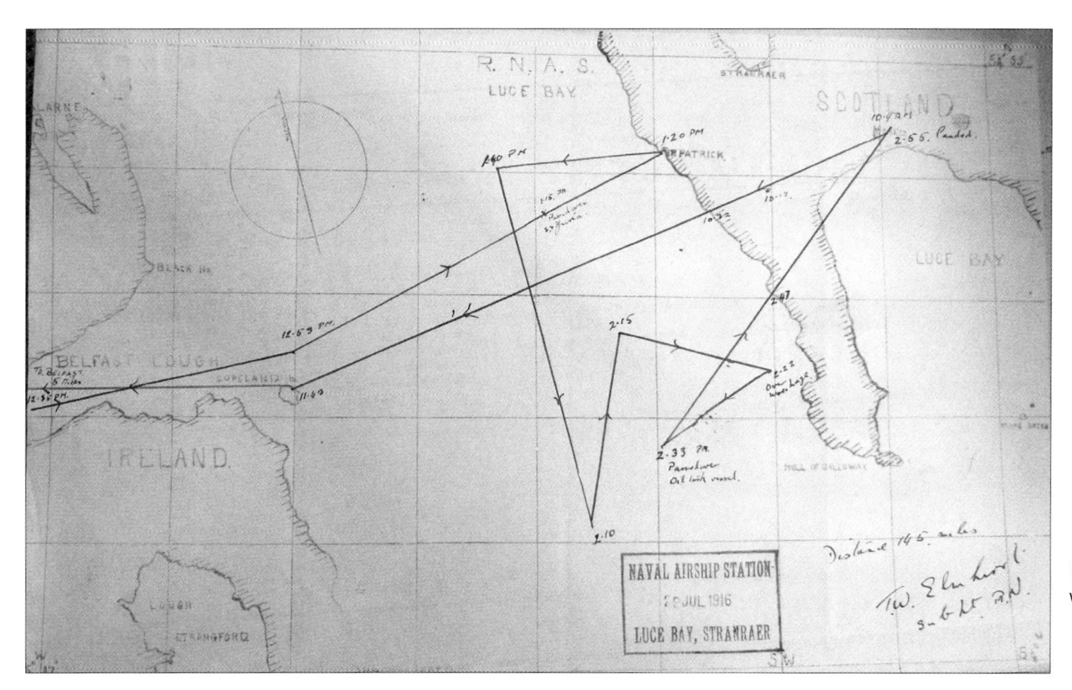

A patrol map from July 1916 bearing Tom Elmhirst's signature. *(National Archive)*

SS20 landing at Luce Bay with the engine stopped. On the ground handlers rush to catch the ropes. *(Elmhirst Family)*

Point 'dropped two bombs at a floating target.' On 29th July Tom Elmhirst in SS20, 'Over entrance to Belfast dock. Observed HMS *Terror* and RMS *Aquitania*.' The *Aquitania* was the third in Cunard's 'great trio', the others being the *Lusitania* and the *Mauritania*.

One of the other pilots went over to Belfast on the same day to see the Senior Naval Officer, Captain Henry Wilkin DSO, and Captain Charles Bruton of *Terror*, a brand-new, 12-gun Erebus class monitor. The intention was to arrange firing the following week, presumably with the ship's gunnery being observed from an airship. It would appear that this did not take place due to adverse weather. *Terror* was one of 30 warships and 44 merchant vessels constructed in Belfast between 1914 and 1918, as well as refitting and repairing many more ships and submarines for the RN. Wilkin served at Belfast as SNO from 1915 to 1919.

Sometimes an airship on early patrol would return with a supply of fish for breakfast. The pilot would maintain station over a fishing vessel and let down a basket containing tobacco or newspapers on the end of the wireless aerial. The fishermen would trade these for a welcome dietary supplement. On one occasion a pilot was returning with a laden basket when fog descended on the landing ground. In the gloom he managed to scrape the roof of the CO's house with the result that the fishy cargo was spilt over the side. Quite how the CO reacted to this 'manna from heaven' landing in his garden has not been recorded.

On 2nd September FSL H Hall in SS23 passed over Stranraer Harbour and saw the destroyer

A group of Luce Bay officers in swimming attire. *(Elmhirst Family)*

A bathing party setting out from Luce Bay in the station's 3-ton Peerless 'liberty-boat', one of 12,000 imported from Cleveland, between 1915 and 1918. *(Elmhirst Family)*

HMS *Thorn*, Lieutenant Douglas Jeffrey RNR, moored at the pier. The other destroyer based at Larne at that time was HMS *Garry*, Lieutenant Henry Binmore. On the same day FSL Hemsley in SS23, having already flown for two and a half hours that afternoon when he had 'observed a Norwegian schooner and two tramp steamers', left the ground at 5.50 pm, returning from patrol to Larne at 9.00 pm and 'landed practically in the dark, directed by Very's lights to Aerodrome.' More night flying took place later in the month.

In November 1916 Commodore Larne made an official request for another armed yacht to supplement *Zara*, so great was the volume of traffic in the North Channel.

On 16th December it was reported that SS20 had to return to base two and a half hours into a patrol, 'owing to the pilot getting cramp in the leg caused by the severe cold.' Then on 27th December Flight Commander IHB Hartford, accompanied by his W/T operator, Air Mechanic Neville, in SS20, 'observed 5 miles SW of Portpatrick Norwegian steamer SS *Hortense Lea*, a collier – empty [hatches off] steaming northward. Circled over ship for examination and reported same to Commodore, Larne.'

Flying the SS Class

It is hard to imagine now just what it would have been like for the crews of the Submarine Scout airships, suspended in an open cockpit, between a few hundred and a few thousand feet above the cold, grey sea, making slow headway against the wind. Luckily two such pilots recorded their impressions. Air Marshal Sir Thomas Elmhirst recalled his experiences as a young pilot of only 19 years of age:

> 'I controlled height by means of a wheel in my right hand linked by wires to elevator

S. 1562. (Established June, 1916.)

SECRET.

REPORT OF OBSERVER.

No. of Reconnaissance	Date	Airship. No.	Airship. Type	Squadron	Pilot	Observer	Ref. Map	Hour at which reconnaissance commenced	Hour at which reconnaissance concluded
6.	27.12.16	20.	S.S.	Luce Bay.	Flight Commander I.H.B.Hartford, R.N.	A.M. Neville, W/T Operator.	X.124.	10.54 a.m.	12.26 p.m.

Time	Place	Observations and weather conditions
10.54 a.m.	Left Air Station, Luce Bay.	Set course W. by N. Magnetic, and as necessary to a position 8 miles E. of the Maidens (vide chart attached)
11.35 a.m.	8 miles E. of Maidens Rock.	A/c S.E. (Mag), and as necessary for Mull of Galloway.
Noon.	Observed 5 miles S.W. of Portpatrick Norwegian Steamer "Hortense Lea" a collier – empty (hatches off) steaming northward. Circled over ship for examination and reported same to Commodore, Larne.	
12.5 p.m.	Observed steamer (British by build and appearance) 7 miles S.W. of Portpatrick. Circled for observation.	
12.26 p.m.	Landed at Base.	

Signed I.H.B. Hartford Observer Flight Commander R.N.

Stn. 7697/16.

A patrol report by Irving Hartford from December 1916. *(National Archive)*

planes stuck on the after end of the gasbag and direction by foot pedals again connected by wires to a rudder at the after end of the gasbag. My other controls, to be operated by the left hand, were the engine, the gas valve, two ballonet air valves and an air pressure control cable. I had a red cord to rip the top off the gasbag in case of a forced landing and a handpump to top up the main fuel tank under my seat from the gravity feed tank for the engine. I had made a wooden bomb sight – quite simply two nails as foresight and backsight, the foresight being moveable to a scale marked with the speed of approach, 20 to 50 knots. My other instruments were mounted on a board to the front of the cockpit – a watch, air-speed and height indicators, an engine revolution counter, an inclinometer, gas, oil and petrol pressure gauges and a glass petrol level indicator. These all had to be monitored closely. Provision was made for illuminating the instruments by four small bulbs. Navigation was by means of chart and floor-mounted compass. A map case with a celluloid front formed the door to a small cupboard.'

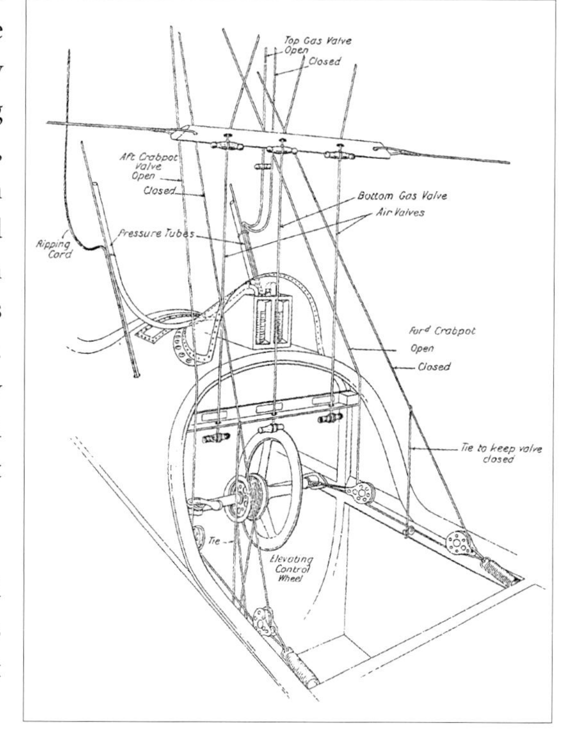

The flying controls of the SS class airship. *(By permission of Patrick Abbott)*

Whilst doing all of this he also had to pass messages back and forth to his wireless operator, read his maps, take compass bearings, plot his course and at the same time keep a constant watch on the sea below.

The wireless operator had to transmit the airship's call-sign every hour to enable two or more of the Marconi wireless stations set up around the coast to obtain a bearing and so plot its position. The stations in Ireland were located at Larne, Kirkistown and Skerries (Drogheda). A classified letter from the CO at Luce Bay to the Commander-in-Chief, Coast of Scotland at Rosyth confirmed the accuracy of this method, 'which has given confidence to pilots when on night patrol or in thick weather in future', though in many cases it was not required as the airships were able to maintain sight of land as they patrolled. Letter and number combinations in Morse code were used to swiftly express certain messages eg. S2P – submarine[s] at anchor off. S3L – submarine[s] observed at [position] steering [course]. S6Z – mines observed at [position]. These were to be used in conjunction with a squared map.

It should be noted that conditions were cramped and confined on board, exposed to the cold and at the mercy of the elements – the speed through the air of these craft could be reduced to only a few knots when flying into a head-wind. This would not be a pleasant experience when returning from a long patrol – tired, hungry and cold. Another airship pilot of the period, TB Williams, wrote:

> 'As the pilot could not leave his little bucket seat during a flight of often many hours duration, he just didn't get a meal. It was also difficult to answer the call of nature. I evolved an arrangement made up from a petrol funnel to which was attached a piece of rubber hose passing to a water tight junction in the hull under my seat. The petrol funnel was hung on a brass cup hook near my elbow. I had some difficulty in inventing a purpose for this gadget when explaining the instruments and controls to the wife of a VIP on one occasion.'

Flight Commander Colmore with the Luce Bay mascot. *(Elmhirst Family)*

The Commanding Officer at Luce Bay, Flight Commander GC Colmore, must have visited Whitehead from time to time, as in September 1916 his name was added to the list of authorised signatures for cheques at the local bank. George Colmore had gained his Royal Aero Club Aviators' Certificate on 21st June 1910, being awarded No 15. Colmore handed over command of Luce Bay to Flight Commander IHB Hartford on 1st November 1916, whose name was duly inscribed in the Northern Bank's ledger five days later.

As the year wore on into the autumn and winter, all was fairly quiet with enemy action being confined to an area around the Isle of Man. Life at Luce Bay was rather pleasant, discipline was not oppressive. When flying was not possible owing to the Scottish climate, personnel enjoyed the navy tradition of a 'make and mend' day. Huts for the men were exceptionally comfortable with washing facilities and a bathroom, rather than a communal washhouse. The officers and their servants enjoyed being billeted in a nearby requisitioned country residence, Dunragit House. However, much was to change in the next few months.

A faded image of seven airship officers and their motorbikes at Luce Bay. from L-R, Hemsley, Turnour, Cole-Hamilton, Elmhirst, Wearham, English and Bryant. *(Elmhirst Family)*

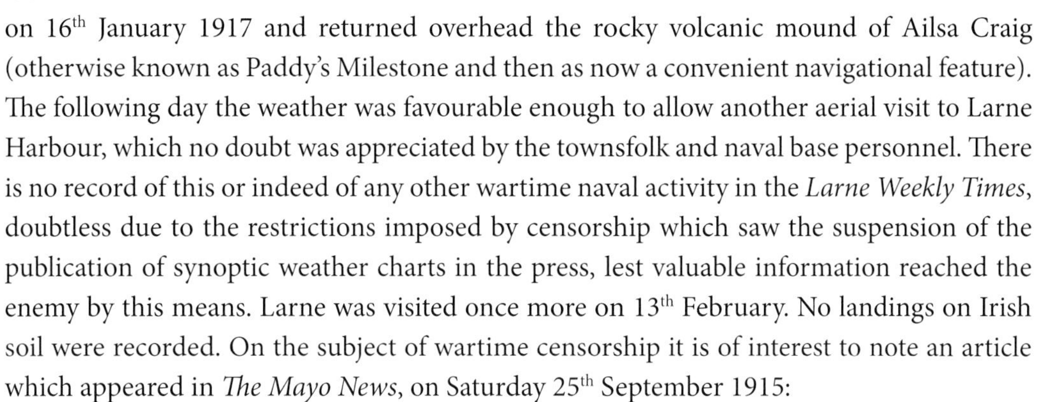

The log of SS20 records that Flight Commander Hartford flew as far as Larne on 16th January 1917 and returned overhead the rocky volcanic mound of Ailsa Craig (otherwise known as Paddy's Milestone and then as now a convenient navigational feature). The following day the weather was favourable enough to allow another aerial visit to Larne Harbour, which no doubt was appreciated by the townsfolk and naval base personnel. There is no record of this or indeed of any other wartime naval activity in the *Larne Weekly Times*, doubtless due to the restrictions imposed by censorship which saw the suspension of the publication of synoptic weather charts in the press, lest valuable information reached the enemy by this means. Larne was visited once more on 13th February. No landings on Irish soil were recorded. On the subject of wartime censorship it is of interest to note an article which appeared in *The Mayo News*, on Saturday 25th September 1915:

> 'An important proclamation signed by Major General LB Friend, Commanding the Troops in Ireland, under date 6th September 1915, has been issued and deals with sketching or photography on the sea shore. It states that it shall be an offence under the Defence of the Realm Act to carry or use a camera or sketching materials within, amongst others, the following area – the coastline for ten miles on each side of Galway City [Auxiliary Patrol Area XX]; the coastline from Roundstone to Clifden, the vicinity of the wireless station at Clifden; the vicinity of the wireless station at Letterfrack; and the coastline from Erris Head to Achill Head, including the whole coast of Blacksod Bay [Auxiliary Patrol Area XIX]. Photography and sketching are permitted in other parts of Mayo and Galway except those places in which troops are quartered or at which military or naval works of any kind exist.'

Chapter 5
The U-boat War Intensifies

THE REQUEST FOR AN extra patrol vessel was answered in January when HMY *Albion III* arrived in Larne, under the command of Captain AH Jones, who would assume the position of senior officer at sea. She was a substantial craft of more than 1000 tons, armed with two 12-pounders and one 6-pounder.

New COs for *Zara* and *Thorn* were appointed, Lieutenant Commander Burn RNR and, on a temporary basis, Lieutenant Frederick Dalrymple-Hamilton. On 24th January, 1917, Commodore (Second Class) Robert Dalton Stevenson Cuming RNR was appointed to HMS *Thetis* in charge of Auxiliary Patrol Area XVII. He was another retired Admiral in his mid-sixties.

The following day the armed merchant cruiser, HMS *Laurentic*, Captain Reginald Norton, struck two mines off Lough Swilly and sank within the hour. She had left Liverpool on 23rd January 1917 bound for Halifax, Nova Scotia with a consignment of gold bullion to the value of £5,000,000 as payment for munitions. At 5.55 pm on 25th January, she was struck a mine on the starboard side and twenty seconds later hit another mine on the same side. The crew of 722 officers and men started to abandon ship and fifteen lifeboats cleared the sides, subsequently only seven of these were saved. 354 men were lost in this sinking in 23 fathoms of water. The minefield had been laid by U-80, Kapitänleutnant Alfred von Glasenapp, one of a class of ocean minelaying submarines.

The wreck of the *Laurentic* was located in 20 fathoms by mine-sweeping trawlers out of Buncrana. Six weeks after the sinking, an operation to recover the gold bullion was mounted and in a series of 5000 dives from the salvage ship, HMS *Racer*, a total of 3186 gold bars

Below: HMY *Albion III* *(Author's Collection)*

Below right: SS *Laurentic* *(Tower Museum Derry-Londonderry)*

U-80 *(Author's Collection)*

The Mayor of Londonderry provided lunch for the survivors the Guildhall. On 27th January 2017 the Mayor of Londonderry hosted a special commemorative lunch in the Guildhall, which the author attended. *(Tower Museum Derry-Londonderry)*

valued at £4,958,708 were retrieved. The cost of the operation was just £128,000. In 2008 the ten-ton deck gun from *Laurentic* was raised from the sea by a team of divers from Co Donegal. They restored the weapon and put it on show at Dowing's pier. This was one of eight six-inch guns mounted on the *Laurentic* and the second to be recovered from the sea bed.

Further reinforcement arrived on 26th February in the shape of HMY *Monsoon*, a two-masted, steel, screw steamer, which had the original name of *Latharna*. Interestingly the name Larne is derived from the Gaelic Latharna.

HMY *Monsoon* *(via Hugh Darlington)*

Unrestricted submarine warfare recommences

Early in 1917 the Royal Navy laid a small field of 72 mines off Belfast Lough. On 1st February 1917 the Imperial German Government declared the resumption of unrestricted submarine warfare. U-boats were available in quantity and this time not only allied shipping but also neutral vessels (such as those on the US register) would be sunk on sight in the eastern Atlantic and the approaches to British ports. This was an enormous gamble, America would inevitably enter the war (the USA declared war on Germany on 6th April 1917) but could the Allies be starved into submission before sufficient reinforcements and materiel could arrive to break the deadlock on the Western Front?

Theobald von Bethmann Hollweg *(Creative Commons Share Alike 3.0)*

The Chancellor, Theobald von Bethmann-Hollweg, assessed the situation facing the Germans:

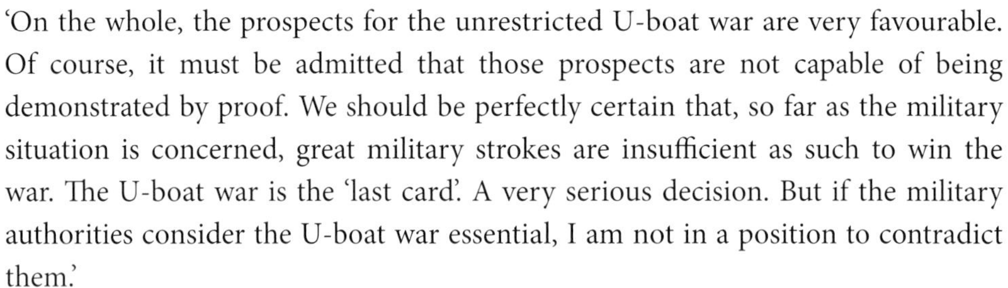
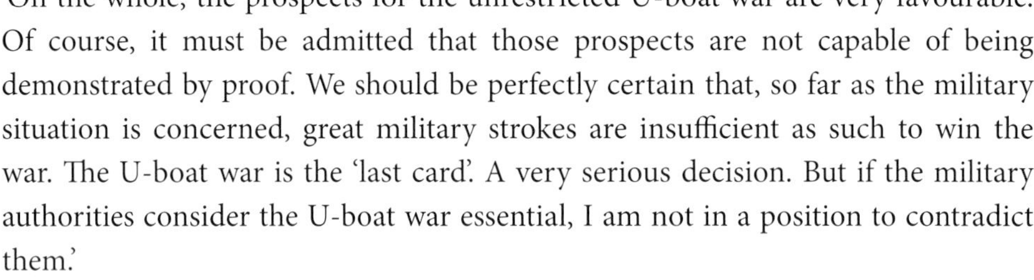

> 'On the whole, the prospects for the unrestricted U-boat war are very favourable. Of course, it must be admitted that those prospects are not capable of being demonstrated by proof. We should be perfectly certain that, so far as the military situation is concerned, great military strokes are insufficient as such to win the war. The U-boat war is the 'last card'. A very serious decision. But if the military authorities consider the U-boat war essential, I am not in a position to contradict them.'

Winston Churchill *(Author's Collection)*

The initial impact cannot be understated. In the first three months, over 1000 merchant ships were sunk. The situation was perilous. Winston Churchill later wrote:

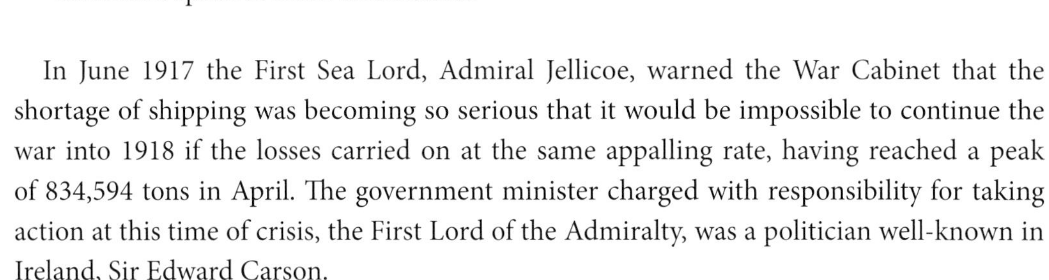

> 'At first sight all seemed to favour the Germans. Two hundred U-boats each possessing between three and four weeks' radius of action, each capable of sinking with torpedo, gunfire or bomb, four or five vessels in a single day, beset the approaches to our islands along which passed in and out every week several thousand merchant vessels. Of all the tasks ever set to a Navy none could have appeared more baffling than that of protecting this enormous traffic and groping deep below the surface of the sea for the deadly elusive foe. It was in fact a game of Blind Man's Buff in an unlimited space of three dimensions.'

Sir Edward Carson *(NIWM)*

In June 1917 the First Sea Lord, Admiral Jellicoe, warned the War Cabinet that the shortage of shipping was becoming so serious that it would be impossible to continue the war into 1918 if the losses carried on at the same appalling rate, having reached a peak of 834,594 tons in April. The government minister charged with responsibility for taking action at this time of crisis, the First Lord of the Admiralty, was a politician well-known in Ireland, Sir Edward Carson.

Much debate ensued within the Admiralty as to the most appropriate type of countermeasures. In the end it was realised that grouping merchant ships into convoys escorted by naval vessels and naval airships was the most effective form of protection. As

RH Gibson and Maurice Prendergast wrote:

> 'Convoys: massing merchant vessels for mutual defence, smaller combatant vessels would now be escorts rather than being frittered away on patrol work. Instead of looking for U-boats, the U-boats would have to come to them if they wished to attack a convoy, 'the most probable point of enemy arrival', as advocated by the strategist and historian Rear Admiral Alfred Thayer Mahan. The submarines would be forced to expend torpedoes rather than the hitherto preferred form of destruction, gunfire and scuttling by bomb.'

This was put into effect from May 1917. Chatterton adds a historical perspective:

> 'Just as in classical times the Roman Fleet convoyed their corn-ships, or in the Middle Ages the Hanseatic League sailed their merchantmen under protection against pirates, or in the sixteenth century the Spanish treasure-ships were escorted from the West Indies, or the same principle was effectually tried during the Anglo-Dutch wars of the seventeenth, and the Anglo-French wars of the eighteenth and nineteenth centuries, so at long last, in spite of some prejudicial opposition on the part of certain master mariners who naturally hated the idea of station-keeping and being herded in fleets, the convoy system now inevitably had to be accepted.'

A torpedoed oil tanker on the point of sinking. *(Author's Collection)*

Winston Churchill added, 'The size of the sea is so vast that the difference between the size of a convoy and the size of a single ship shrinks in comparison almost to insignificance. There was in fact very nearly as good a chance of a convoy of forty ships in close order slipping unperceived between the patrolling U-boats as there was for one single ship; and each time this happened, forty ships escaped instead of one.'

A convoy in the Irish Sea *(National Archive)*

An early UC Class U-boat. *(Author's Collection)*

Thorn had a most unusual escort duty on 26th March, accompanying the Imperial Russian Navy's cruiser *Varyag* which was on its way to Cammell Laird in Liverpool for refit.

Two flotillas totalling 14 RN submarines were established in Ireland in March 1917, under the command of Captain Martin Nasmith VC. The first patrol was made on 5th March when six submarines sailed from Queenstown to an area west of Tory Island, escorted by HMS *Adventure*. The *Platypus* and *Vulcan* flotillas, named after the submarine depot ships HMS *Platypus* and HMS *Vulcan*, provided supplies, mobile workshops, and accommodation to resting submariners. These flotillas were mobile, and at various stages were based in Queenstown, Berehaven (both in Co Cork), Lough Swilly and Killybegs (both in Co Donegal), with the former to the north and the latter to the west. In April 1917 the initial deployments were the *Vulcan* flotilla to Buncrana and the *Platypus* flotilla to Killybegs.

The submarine depot ship HMS *Vulcan*. *(NMRN)*

A contemporary postcard of UC-5. *(Author's Collection)*

A mine being loaded into a UC Class U-boat. *(Author's Collection)*

As the year progressed UC type U-boats laid mines off the north coast of Ireland, while hydrophone drifters worked from the Mull of Galloway across to south of Belfast.

The largest number of U-boats engaged in operations at any one time is thought to be 61, in June 1917, based on an active strength of 200 and allowing margins for replacement of losses, docking, training, replenishment and moving to and from war stations.

On 1st May UC-65 stopped the SS *Helen*, which was on her way to Bangor with a cargo of coal, near the Copeland Islands and it was sunk by using a small explosive charge. The crew had to row to Donaghadee and raise the alarm. There is a story, which is unlikely to be true, that the U-boat commander, Kapitänleutnant Otto Steinbrinck, advised the crews of the local bus times as they made for the shore. The following day four coasters were sunk by UC-65 off Ballyhalbert, the *Derrymore*, *Morion*, *Saint Mungo* and *Amber*, as well as the wooden three-masted schooner, *Earnest*, off Portavogie. This last attack is of particular interest as Steinbrinck disguised his mine-layer as a sailing boat. Moreover, he presented Captain James Ferguson, *Earnest*'s Master, with his binoculars by way of compensation. By this time the alarm had been raised and the U-boat came under attack by the armed trawler, *Goshawk II*. Once in range, the trawler fired shells at the U-Boat but it crash dived and headed towards Ardglass. At noon Steinbrinck said that four depth charges were dropped from the air and later he said that a small airship was leading the hunt for his submarine.

Kapitänleutnant Otto Steinbrinck *(Author's Collection)*

On 26th June the first US troops landed in France, the six transports having been safely

HM Submarine D-3 under way. *(NMRN)*

convoyed across the Atlantic and met by six US destroyers out of Queenstown for the final stages of their voyage. In July 1917, prior to the arrival of US submarines, Commander Boyd USN, sailed on patrol with D-3 from Lough Swilly, to observe methods and tactics.

Developments at Bentra and Larne

During the course of a Western Patrol, Flight Commander Hartford landed at Whitehead twice on 7th April 1917, flying SS20, at 1.00 pm and again at 3.00 pm. FSL AV Pullan made a similar stop on 3rd May and again on 12th in SS23. The facilities there were being improved to enable the mooring out station to make a greater contribution in the light of the highly increased threat. Ditches were filled in. A portable airship shed was erected, consisting

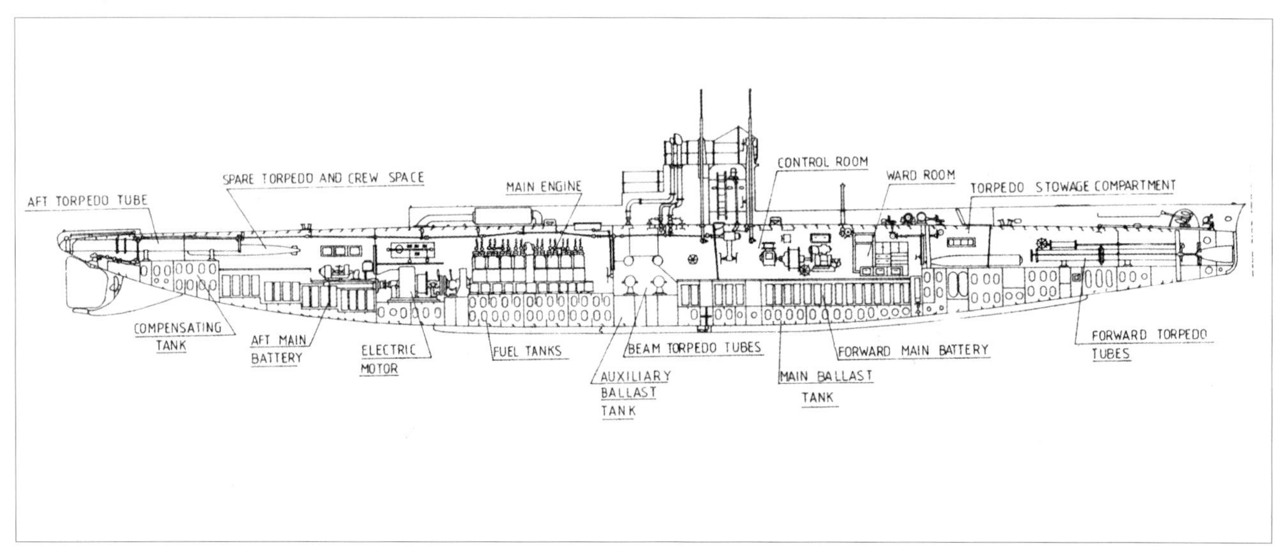

Layout of a typical RN submarine. (Author's Collection)

The ground crew pose for a photograph in front of the shed. *(D&N Calwell Collection)*

of a steel frame covered by canvas, measuring 150 feet long by 45 feet wide and 50 feet high. (It will be recalled that official approval for the provision of a small airship shed had first been given in July 1915!) Wooden huts on brick and concrete foundations were built to act as accommodation on site. Rations for the men stationed at Bentra were provided by the Army Service Corps, Belfast, via the 6th Royal Fusiliers at Carrickfergus Castle. It was noted that the disposal of 'by-products' was not working satisfactorily, 'owing to the delay in receiving the necessary receptacles from the contractor.'

Flt Cdr IHB Hartford at Bentra. Note the huts in the background. *(D&N Calwell Collection)*

The primary task for the airship making its way to Bentra early each morning was to escort the *Princess Maud*. Around midday, when the prevailing wind was often unfavourable, the airship could be housed, ready to take to the air again in the evening when the wind dropped. The introduction of the convoy system brought fresh duties, scouting for submarines on the surface or the wake of a periscope. Co-operation between naval airmen and the warships below was of the highest importance. Every hour the airship would transmit its call sign and so allow its position to be plotted by cross bearings from strategically placed wireless stations. If anything suspicious was sighted then warships could be directed to the precise location without delay. An airship could take position to windward of a convoy and could swoop down fairly quickly to investigate a possible threat at a speed nearly twice that which a destroyer could achieve. It could receive messages by Aldis lamp from escorting warships and it could scout ahead for mines – which could be detonated by machine-gun fire.

Also in April Commodore Larne requested that his destroyers and yachts should be

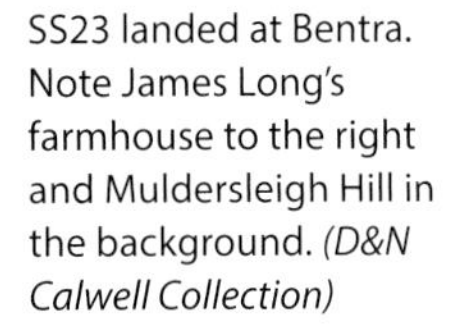

SS23 landed at Bentra. Note James Long's farmhouse to the right and Muldersleigh Hill in the background. *(D&N Calwell Collection)*

equipped with improved directional hydrophones as a matter of urgency. On 6th May *Zara* investigated a hydrophone contact and attacked with depth charges, as did an accompanying armed trawler.

The first airship to use the shed at Whitehead was SS20 on 5th June, flown by FSL WEC Parry. He encountered some rough weather while on a routine flight and had made for shelter. He had to stay there for four days until the conditions improved, returning to Luce Bay after a Western Patrol on 9th June. That evening Flight Commander Hartford made another three hour Western Patrol, landing at Whitehead to pick up Air Mechanic Futcher. On 10th June SS20 returned to Whitehead, flown by FSL R Chambers and spent eight hours on the ground there between patrols. Soon afterwards Pullan in SS23 experienced engine trouble and was able to find a safe haven there as well. On 11th June Flight Commander Hartford flew to Bentra, bringing with him a mechanic to repair SS23's engine.

Larne Naval Base received further augmentation on 16th June when a pair of paddle minesweepers, No 554 *Her Majesty* and No 555 *Princess Beatrice* arrived for sweeping duties in Belfast Lough. They were soon joined by No 841 *Lady Clare* and No 842 *Belle*, all under the command of the Officer-in-Charge Minesweeping, Lieutenant EE Balding RNR .

They were 'mercantile conversions to take advantage of shallow draught, high speed and manoeuvrability.' *Thorn* received a new, permanent CO, Lieutenant Commander Claude Hamilton RNR.

An indication of the increasing range of aerial wireless communication may be gained from this report from Flight Commander Hartford:

> 'On 30th June 1917 Chief Petty Officer W Martin, W/T Operator, was flying in SS20 with Flight Sub Lieutenant R Chambers as pilot. Using a Sterling transmitter, he was able to send message from Whitehead in County Antrim to Anglesey Naval Airship Station, in daylight. The distance covered was approximately 100 miles. It is believed

A paddle minesweeper. *(Author's Collection)*

that this performance will be of interest to wireless operators at other Stations.'

He also noted that in the past month 47,050 cubic feet of hydrogen had been manufactured on-site, using caustic soda and silicol. The process was as follows: Powdered ferrosilicon was fed in a controlled manner into a closed stirred tank containing a very hot, strong solution of caustic soda. The ferrosilicon reacted rapidly with the caustic soda, producing a mixture of steam and 99% pure hydrogen. Usually the gas was then passed through cold water in a 'scrubber' to condense the steam, collect the hydrogen and remove some of the (poisonous) impurities. The chemical reaction produced a lot of heat, which was used to keep the temperature of the tank at around 115 degrees Celsius. A residue consisting mainly of sodium silicate (water glass) was left in the liquid in the tank. When the batch of caustic soda had become exhausted, the tank had to be drained as soon as possible, otherwise the residue would solidify in the tank. Inflating SSZ11 took only 35 minutes. Internal padding in the airship shed had been completed, fire hydrants were being installed and a contract had been let for the disposal of silicol sludge. A Station garden had been created, where the 'ship's company' grew their own fresh produce.

In early July Flight Commander Hartford came across once more to Bentra with an engineer to fix SS23's troublesome engine. However, changes were afoot – SS20 and SS35 had been inspected by the Admiralty Board of Survey and condemned as being unfit for service. On 10th July SS23 was, 'dispatched to Whitehead Naval Airship Base, in order to make room for rigging the new ships.' On the following day SS24 had flown up from RNAS Anglesey to take on some of the burden but a much greater augmentation was at hand – the first of the new SSZ Class to arrive at Luce Bay by rail to Dunragit for inflation and rigging. Before the end of July both SSZ11 and SSZ12 had made their first operational patrols from Luce Bay. They were joined by SSZ13 in August. Of the older SS type only SS23 at Whitehead remained operational.

SS23 is prepared for flight from Bentra. *(D&N Calwell Collection)*

SS20 in the airship shed at Bentra. This photograph was taken on a long exposure which creates the illusion of an eight-bladed propellor. *(D&N Calwell Collection)*

Not everything seen from the air was easy to identify. Airships were reported sighting and chasing a submarine near Belfast Lough on 4th August, on investigation it was found to be a marker buoy!

Q-ships

A Q-ship was lost on 5th August. The *Chagford*, Lieutenant DG Jeffrey, had departed from Buncrana three days before and was searching for U-boats 120 miles north-west of Tory Island. She

Kapitanleutnant Wolfgang Steinbauer. *(Author's Collection)*

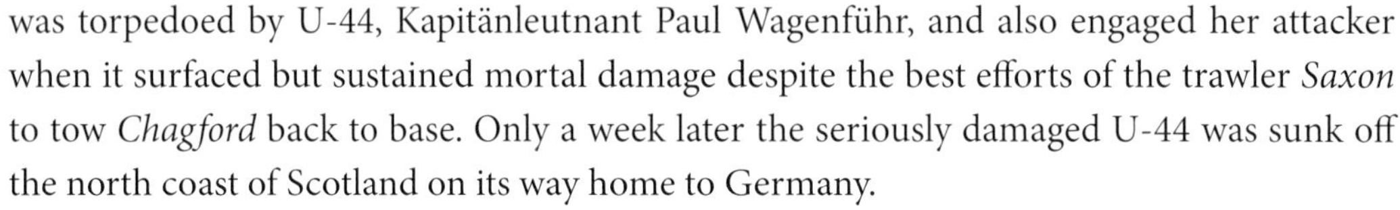

was torpedoed by U-44, Kapitänleutnant Paul Wagenführ, and also engaged her attacker when it surfaced but sustained mortal damage despite the best efforts of the trawler *Saxon* to tow *Chagford* back to base. Only a week later the seriously damaged U-44 was sunk off the north coast of Scotland on its way home to Germany.

On 15th August 1917 Lieutenant Commander William Richardson in the submarine D-6 was cruising in company with the top-sail, three-masted schooner HMS *Prize*, Lieutenant Commander William Sanders VC, DSO, a Decoy or Special Service Ship, well-armed with concealed guns, 150 miles north-west of Rathlin Island. UB-48 was commanded by Kapitänleutnant Wolfgang Steinbauer, who Chatterton describes as, 'a young man, full of daring, smart, cold as steel, efficient, quite sure of his own ability, dignified, perhaps even somewhat self-complacement, yet a fine naval officer.' Steinbauer was unaware that

Both *Result* (above left, *Courtesy Carrickfergus Museum*) and *Mary B Mitchell* (above right, *Courtesy Graham Walton*) were constructed at Paul Rodger's shipyard in Carrickfergus and were also employed as Q-ships.

the innocent-looking sailing boat was a RN warship, so closed in on the surface while Richardson, about three-quarters of a mile astern, dived for a counter-attack. Suddenly *Prize* hauled down her decoy Swedish flag, hoisted the White Ensign, and opened fire on the U-boat. Heavy seas prevented D-6 from getting close enough to launch her torpedoes and, having been struck by two shells, UB-48 quickly submerged to seek safety in the depths.

But this was not the end of the story. Now aware of the true identity of the little schooner the U-boat captain followed her to the north-west and, early on the morning of 14th, a well-placed torpedo blew her to pieces. D-6 was following two miles astern and the men on her bridge saw the explosion but, although they hurried to the scene, not a single member of the Q-ship's crew survived. Sanders had won his VC in *Prize* for a remarkable action on 30th April off the south west coast of Ireland. U-93 was disabled, its CO, Kapitänleutnant Freiherr Edgar von Spiegel von und zu Peckelsheim, was knocked into the water and captured but the submarine managed to limp home to Germany 10 days later.

Prize was towed into Kinsale Harbour by HMML 161, Lieutenant Hannah, where the wounded were disembarked.

Lieutenant Commander William Sanders. *(Author's Collection)*

Kapitänleutnant Freiherr Edgar von Spiegel von und zu Peckelsheim of U-93. *(Author's Collection)*

Chapter 6

Enter the SSZ Class

THE SS ZERO WAS built to the design of three RNAS officers at RNAS Capel, near Folkestone, Commander AD Cunningham, Lieutenant FM Rope and Warrant Officer Righton. The car was specifically designed to be streamlined in shape and was constructed almost like a boat, with a keel and ribs of wood with curved longitudinal members. The whole frame was braced with piano wire and then floored from end to end. It was enclosed with 8-ply wood covered with aluminium. The crew of three consisted of the wireless telegraphist/observer/gunner in the front, with the pilot in the middle and the engineer in his own compartment to the rear. A machine-gun could be mounted either to port or starboard, operated from the front seat and two 110 lb bombs could be carried. The car, as well as being boat-shaped, was watertight so the airship could land on calm water. It was powered by a 75 hp Rolls-Royce Hawk six-cylinder, vertical in-line, water-cooled engine driving a four-bladed pusher propeller. 200 of these were manufactured under licence by Brazil Straker of Bristol, which was the only company entrusted by Henry Royce to build complete engines. It was a superb creation and was test run for the first time at the end of 1915. The words of an airship pilot tell it all, 'The sweetest engine ever run – it only stops when switched off or out of petrol.' It gave the airship a top speed of 53 mph and a rate of climb of 1200 feet per minute. Slung on either side of the gasbag were two petrol tanks made from aluminium. The gasbag had a capacity of 70,000 cubic feet. It was of the same

An SSZ Class airship is being hauled down (in this case at RNAS Pembroke in 1917). The trail-rope is in the hands of the handling party, the engine is stopped and the engineer is standing while switching everything off. *(Ces Mowthorpe Collection)*

The car of the SSZ Class airship was streamlined and boat-shaped. *(Rolls-Royce Heritage Trust)*

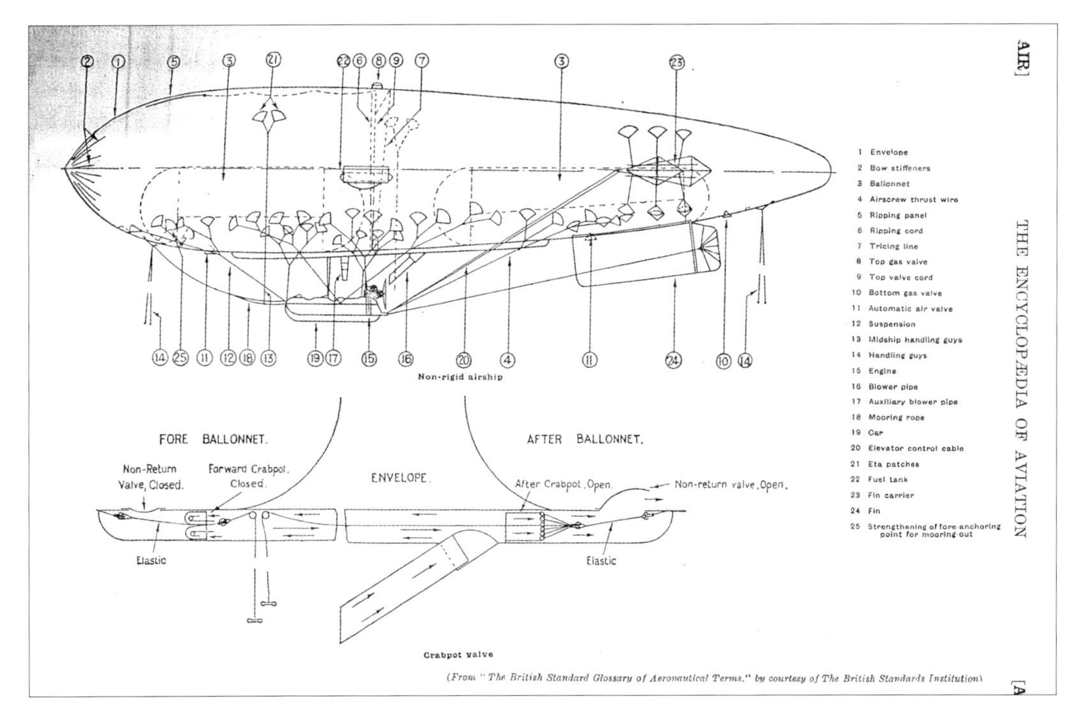

A plan drawing of the SSZ Class airship. *(Author's Collection)*

length as the SS Class gasbag but was of a slightly greater diameter. The nose of the gasbag was reinforced by radially positioned canes to prevent it buckling at speed. As with the SS Class it was attached to the car with cables secured to the envelope by kidney-shaped 'Eta' adhesive patches, which were also sown on, so spreading the load evenly. The SSZs or Zeros as they were known, were more stable in flight than the SS Class and had much greater endurance. They were able to fly in weather conditions that would have prevented the earlier type from operating. Its unit cost was about £5000.

The arrival of the more capable SSZs and the greater flexibility allowed by the erection of the shed at Whitehead ensured that September was a very busy month with regard to

An SSZ Class airship is prepared for flight. *(Rolls-Royce Heritage Trust)*

patrols, training and test flights. A report submitted by the CO at the end of the month noted that with the onset of autumn, rain, high winds and dull, threatening clouds had been prevalent. The same treacherous wind characteristic as the previous year was a familiar hazard, 'the wind hauling from NW to S and increasing rapidly.'

A table summarised the performance of each airship during the month.

Airship	Time in the Air	Miles Flown	Petrol Consumed	Oil Used
SSZ12	57 hours 25 minutes	1722	248 galls	5¼ galls
SSZ13	37 hours 38 minutes	1129	180 galls	4¼ galls
SSZ20	40 minutes	20	55 galls (filled up)	4 galls
SS23	10 hours 53 minutes	327	84 galls	12¾ galls
Total	**106 hours 36 minutes**	**3198**	**568 galls**	**26¼ galls**

In spite of the conditions a 'fair amount' of patrol and escort work was carried out, a highlight of which was a patrol of nine hours and five minutes by FSL Noel Grabowsky in SSZ12, 'for three hours during this flight the airship was flying full out and only just holding her own against the wind, but by using a very skillful knowledge of the local windage, the pilot was able to pick up a northerly slant which allowed him to land at the airship base.'

Submarine verses U-boat

On 12th September, U-45, Kapitänleutnant Erich Sittenfeld, was destroyed near Rathlin by the British submarine D-7, which fired a single torpedo from a range of 800 yards and whose commander, Lieutenant Oswald Hallifax, was awarded the DSO. Engine Room Artificer RJ Ballantyne received the DSM. Hallifax later wrote, 'On 10th September we received a signal ordering D-7 and D-8 to take up position between the Hebrides and the north Irish coast. The object of this was to intercept a Hun submarine concentration in the North Channel.' On the morning of 12th D-7 surfaced to charge her batteries:

A postcard image of HM Submarine D-7. *(Martin & Lindy Lovegrove Collection)*

'I was sitting on top of the foremost periscope standard, at

> 10.30 we climbed up on a big roller and as we surmounted the crest, the signalman yelled out, 'Conning tower 2 points on the port bow.' Hallifax manoeuvred his boat into an attacking position but wanted to be absolutely sure that he was going to torpedo a U-boat and not D-8. Having made a positive identification based on the fact that he saw two deck guns rather than one, 'I fired the stern tube… there came to us the sound of an explosion. Everyone cheered and the Chief ERA nearly knocked me out with a clap on the back. Astern of us was a big patch of oil and some objects.'

They picked up two survivors, who, fortunately for them, had been sitting by the conning tower hatch opening having a smoke. Some months earlier Sittenfeld was involved in what would appear to have been a rare humanitarian gesture. The iron four-masted barque *Thor II* was on passage from the remote southern island of South Georgia to Queenstown with a cargo of 1300 barrels of whale oil. It was stopped by U-45 and sunk by torpedo 80 miles off the Irish coast. The crew were given half an hour to get into the life-boats. Then Sittenfeld gave the order to the Master, Captain Jacobsen, to row alongside and bring the chronometer and ship's log. That was done and the Captain, his wife and young daughter were taken aboard the submarine to avoid them having to remain at sea in an open boat. The U-boat then towed the life-boats for four hours in a NE direction but cut the tow and suddenly disappeared beneath the waves. The crewmen then put up their sails, steered towards land and were rescued by a Royal Navy patrol boat. Captain Jacobsen and his family were well treated on board. They were subsequently landed at Heligoland and repatriated to Norway.

D-7 was part of the *Platypus* flotilla. In 1917 four destroyers were based across the lough from Rathmullan at Buncrana; HMS *Medina*, HMS *Orestes*, HMS *Orford*, and HMS *Plucky*. The depot ship there from September 1917 to December 1918 was HMS *Hecla II*, which had

The depot ship HMS *Hecla II*. *(NMRN)*

Below: A convoy seeks cover by making a smoke screen *(Author's Collection)*

Above: A 30-knot 'Turtle-back' destroyer similar to HMS *Wolf*. *(Author's Collection)*

been built in Belfast in 1878 as a torpedo boat carrier.

The destroyer HMS *Wolf*, Lieutenant-in-Command George Parry RNR, joined the Larne-based flotilla in September. Chatterton writes:

> 'By the autumn of 1917, the North Channel was depending rather on hydrophones than on nets. Thus, a line of hydrophone drifters was stretched on patrol across the mouth of the Clyde and another line from the Mull of Galloway to Skulmartin [off Ballywalter]. The convoys of shipping passing through this anxious North Channel were screened by the patrol vessels based on Larne, while motor launches and drifters scoured the district with their hydrophones.'

Excitement at Whitehead

It must have been particularly exciting that summer of 1917 for the children of what is, even today, a quiet country district. What must they have felt seeing a strange craft descending from the sky with its helmeted, goggled and muffled crews? Fortunately an eye-witness to these times has spoken to the author, Mrs Nancy Calwell (back then Nancy Wisnon), who was a young girl six years of age, living near Bentra. Her uncle was James Long, the farmer who owned the land. When the drone of an airship's engine could be heard approaching, the local children would run down the side of the adjacent field to watch through the hedge. A photograph has survived of Nancy and her friends peering over a gate in awe at the sight which they beheld. Soon Nancy was adopted as a mascot by the airmen and given sweets and chocolate. One day in particular she was given a never to be forgotten treat. An airship

SSZ11, landed at Bentra under the watchful gaze of Nancy and her friends. *(D&N Calwell Collection)*

was coming in to land, as usual nose to wind. The pilot threw down a rope for the mooring party of naval personnel, farmers and labourers to grab hold of. When it drew closer to the ground they could reach up for the guy-ropes attached to the bows and the stern. Once it was close enough to the ground, Nancy was lifted into the cockpit onto the pilot's knee. The landing party moved forward tugging on the ropes to 'walk' the airship into the shed (which also saw use as an extra hay barn). Nancy was flying a few feet above the ground and could dream of being an airship pilot ranging out over the sea.

Handling on the ground could be a tricky business as an airship presented a sizeable bulk to the wind and was naturally buoyant in this element. When safely in the shed, maintenance could be carried out by the riggers and mechanics, with their patches, rubber solution and dope. For take-off the airship would be made positively buoyant so that it could be 'walked' out of the shed. Trim would be checked, the engine started, the order to 'Let go' would be given and the craft would rise gently into the wind.

SSZ20 on the ground. *(JM Bruce/GS Leslie Collection)*

U-boat successes

October brought successful U-boat attacks on the armoured cruiser, HMS *Drake*, and the small collier SS *Main* off the Antrim and Galloway coasts respectively. *Drake* was mortally hit by a torpedo from U-79, commanded by Kapitänleutnant Otto Rohrbeck.

The cruiser was about five miles off Rathlin Island when she was hit. The torpedo struck No 2 Boiler Room and caused two of her engine rooms and the boiler room to flood, killing

The armoured cruiser HMS *Drake*. *(Martin & Lindy Lovegrove Collection)*

18 of the crew. These gave her a list and knocked out her steering. Captain Stephen Radcliffe decided to make for Church Bay on Rathlin and on the way accidentally collided with the merchant ship SS *Mendip Range*. The collision did not damage *Drake* much more than that already inflicted, but *Mendip Range* was forced to run aground near Ballycastle. *Drake's* crew was taken off before she capsized later that afternoon.

Main was intercepted by the UC-75, under the command of Oberleutnant zur See Johannes Lohs, east of Luce Bay and sunk by gunfire. On the same day as the attack on *Drake*, 2nd October, the steamer SS *Lugano* sank off Rathlin Island after striking a mine, in all probability laid by U-79. Another mine disabled the destroyer, HMS *Brisk*, Lieutenant Commander Henry V Hudson, which had gone to the aid of *Drake*. Her bows were blown off and her stern section was towed into Londonderry. UC-75 also sank the armed steamer SS *WM Barkley* on 12th October, which was owned by the Guinness Company and was carrying a consignment of the famous stout from Dublin to Liverpool.

Flight Commander Hartford responded with an increased level of aerial activity despite poor weather and gale force winds. On 11th October SSZ12 was escorting an inward bound convoy through the north-west passage. The wind got up into almost a gale when she was off the Irish coast. The speed of the airship by the pitot tube showed 47 knots and in consequence the pilot found that he could make little or no headway. He therefore left the convoy when it was safely through the passage and decided to run for the RFC aerodrome at Turnberry, being assisted to land by RFC personnel and berthed in the lee of the hangars. The pilot had advised base of his intentions by radio and a handling party arrived at Turnberry shortly after the airship. It was deflated and returned to Luce Bay with only minor damage having been sustained and was ready for service again within the week. Flight Commander Hartford was full of praise for the SSZs, noting that a greater amount of patrolling was possible than with the faithful, old SS class.

On 16th Luce Bay received a visit from the 'Director of the American Air Service.' Later in the month Commodore Larne requested that an airship be sent to search for mines reported to be floating off the coast. A fourth SSZ was ready, SSZ20, with the result that SS23 was taken out of service, her last flight being the return from Bentra on 31st October 1917 in the hands of Flight Sub-Lieutenant Pullan. SS23 spent the bulk of the time in her last couple of months of service based at Bentra. On 3rd November UC-75 had further success, torpedoing the SS *Atlantian*, en-route from Galveston to Liverpool, which managed to limp into port, only to be repaired and sunk six months later. The weather in November continued to be poor – on the 25th strong winds gusting to 73 mph tore apart the canvas door curtains of the Whitehead shed.

Both doors open, an airship can clearly be seen inside the shed. *(D&N Calwell Collection)*

Winds of this velocity are tabulated on the Beaufort Scale as Force 11 to 12, which equates to a violent storm or hurricane. There was a danger that the airship sheltering inside would be carried away or badly damaged. A telephone call brought the army to the rescue in horse-drawn wagons. After helping to make everything secure they 'ate up the available rations and departed.' The shed was unusable for nearly a month until a sailmaker could be flown over to effect repairs. The CO noted in his monthly report that the shed at Larne being out of commission, 'materially affected escort and patrol work at Luce Bay, and the importance and the use of the Larne shed was not fully appreciated until its temporary loss was sustained.' Between 24th November and 5th December it was impossible to take the airships out of the shed. A high level of enemy activity continued – helped by often atrocious weather. Reports of sinkings, near misses and running battles between merchant ships and U-boats poured into the operations offices at Luce Bay and in Larne. The airships' level of activity in November was laudable in the circumstances:

Airship	Time in the Air	Miles Flown	Petrol Consumed	Oil Used
SSZ11	42 hours 55 minutes	1290	128 galls	1 gall
SSZ12	26 hours 10 minutes	785	80 galls	8 galls
SSZ13	20 hours 8 minutes	605	80 galls	3¼ galls
SSZ20	30 hours 55 minutes	930	108 galls	¾ gall
Total	**129 hours 9 minutes**	**3610**	**396 galls**	**13 galls**

U-90 was one of six of the U-87 class. *(Author's Collection)*

Three particularly lengthy flights were made, two on 10th December, of eight hours and 15 minutes by FSL SB Harris and of six hours and 35 minutes by FL JAB Ball, while the redoubtable Harris was out again for nine hours and five minutes on 11th. On Thursday 20th December SSZ13, flown by FL WEC Parry, suffered total engine failure in the air from mechanical causes, the first such incident since 16th November 1916. Luckily he was over the aerodrome at a height of 150 feet, as he had turned for home when the engine started giving trouble. He made a slow, controlled descent in a field about a mile away and was greeted by a landing party which had rushed to the scene by 'doubling across country.' The engine failure was found to have been caused by the condenser having burnt out on the magneto. On a more general point the CO reported that, 'Probably the most noticeable feature in flying throughout the past month is the intense cold experienced by the crews of Zero ships in this locality. The surrounding mountains are 2000 feet high, and are entirely snow-capped, which makes the prevalent winds bitterly cold.'

Further sinkings

On Christmas Eve the steamer SS *Daybreak*, which was carrying a cargo of maize from Huelva to the Clyde, was sunk off the coast of Co Down by U-87, Kapitänleutnant Freiherr Rudolf von Speth-Schülzburg.

An eye witness, John Bailie, was on board the South Rock Lightship, 'I remember being on the South Rock as a temporary, 2s/6d a day and feed yourself. About midday the *Daybreak* was torpedoed and 21 were lost. Her nose was cut clean off. It happened so quick her propeller was going round in the air as she sank. You talk about explosions, boilers bursting one after another.'

A few days later, on 27th, the Q-ship, HMS *Lothbury*, Lieutenant Edward Wilkinson DSC, RNR, engaged UC-75 in a brief exchange of gunfire off Burrow Head, the headland separating Luce Bay from Wigtown Bay. The Q-ships would appear as a defenceless merchant ship or fishing vessel but with concealed armament, designed to lure a U-boat within gun range and then open fire. The impression had to be one of an easy prey, whose crew had taken to the boats, and so a 'panic party' had to leave the ship in the lifeboats, leaving the gun crew

The Q-ship HMS *Lothbury. (Author's Collection)*

still concealed behind cover. As the U-boat always had the option of simply torpedoing the Q-ship some of them were filled with timber to increase their chances of staying afloat, in the hope that the U-boat would then surface to finish her off with her deck gun. Admiral Bayly later wryly noted that the first two Q-ships supplied by the Admiralty were, 'too large and laden with coal. The only apparent use of the coal was to enable them to sink at once if torpedoed, and so save the crew from the trouble and discomfort of trying to prolong their lives by staying on board.'

The war came even closer to Whitehead on 28th December 1917 when the Elders & Fyffes liner SS *Chirripo* was sunk by a mine, probably laid by UC-75, about half a mile south-east of the cliffs at Black Head, which was on the far slope of Muldersleigh Hill from Bentra. Two days later the armed trawlers *Angerton* and *Davara* had an inconclusive engagement with a U-boat off The Maidens, islands to the north-east of Larne, exchanging gunfire and dropping depth charges. 1917 ended with four SSZ airships in commission at Luce Bay. The number of personnel at the base had by this time expanded to about 170.

The unfortunate SS *Chirripo*.

Further additions had been made to the destroyer force at Larne in December, HMS *Express*, Lieutenant Commander Thomas Young and HMS *Osprey*, Lieutenant-in-Command Alexander Macrae.

Chapter 7

1918: The climax of the struggle

EARLY IN 1918 SUBMARINES became particularly active off the north coast of Ireland. To protect the approaches to the Clyde, Belfast and Liverpool, a deep mine-field was begun in the North Channel by RN minelayers assisted by the USS *Baltimore*, a veteran cruiser, which had served in the Spanish-American War of 1898, converted into a minelayer.

The entire field was planned to comprise 10,000 mines but was incomplete by the time of the Armistice. Although it took some time to develop a really efficient mine (the successful H2 pattern being introduced in 1916) it rapidly became a most effective weapon against the U-boat, eventually sinking 25 per cent of the total.

The combination of adverse weather and intense enemy action continued unabated. On 5th January 1918 the SS *Knightsgarth* was sunk by U-91, Kapitänleutnant Alfred von Glasenapp, five miles WNW of Bull Point, Rathlin Island, while on passage from Lough Foyle to Barry in Wales with a cargo of coal, tinned food, flour and some small arms. Captain JB Gordon and his crew were assisted to the shore by five local fishermen. For their bravery they each were sent a wallet containing £5 by Rea Shipping Co Ltd. The sinking was reported to Queenstown by Portballintrae Coastguard.

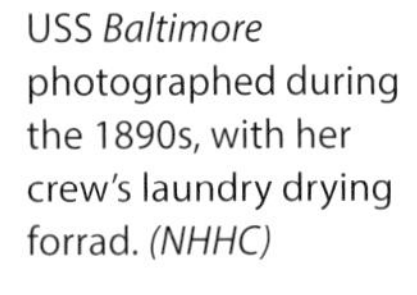

USS *Baltimore* photographed during the 1890s, with her crew's laundry drying forrad. *(NHHC)*

RMS *Andania*.
(Author's Collection)

Then on 27th, the Cunard liner, RMS *Andania*, was sunk two miles NNE from Rathlin by U-46, Kapitänleutnant Leo Hillebrand. The SS *Tuscania* was lost on 5th February, even though it was part of HX-20, a well-escorted convoy; torpedoed by UB-77, Kapitänleutnant Wilhelm Meyer, north of Rathlin.

Of the 2013 American troops and 384 crew, 210 perished, the others being saved by the escorts and brought to Larne and Buncrana. On 12th February U-89, Kapitänleutnant Wilhelm Bauck, was rammed and sunk with all hands by the armoured cruiser HMS *Roxburgh*, which was escorting a convoy around the north coast of Ireland and was

The armoured cruiser HMS *Roxburgh*. *(NMRN)*

commanded by Captain Gerald W Vivian RN.

The SS *Rio Verde* was torpedoed and sunk without warning on 21st February, four miles off the Mull of Galloway, with 20 lives lost, including the Master. On 26th February another ship went down off Black Head, the Anchor Line SS *Tiberia*, which was torpedoed by U-19, a pre-war veteran, Kapitänleutnant Johannes Spiess, who the day before had sunk the tanker, SS *Santa Maria* off Rathlin, which was carrying a cargo of two million gallons of fuel oil. He later wrote:

> 'In March 1918 I had my nicest attack. I was blocking the entrance to the North Channel near Rathlin. The high cliffs with the barns and the lighthouses were a nice picture for a painting. Around noon there were six airships in sight [!], which were flying up and down the cliffs with the wind, just like it was a sport.'

It seems unlikely that six airships were present, which is certainly not corroborated by any official account seen by this author.

A peak was reached in March 1918 when no less than 65 patrols were flown during the month. A new pilot had arrived in February, Flight Sub-Lieutenant AH Crump. Fortunately his Flying Log Book has survived to give an insight into the duties of the airship crews in the final months of the war. After five hours of training in the capable hands of Flight Commander Hartford and Flight Lieutenant Pullan, Crump made his first operational patrol in SSZ20 on 25th February, escorting a convoy for seven hours and 35 minutes.

On 1st March another convoy escort, the large Allan liner and armed merchant cruiser, HMS *Calgarian*, Captain Robert Newton RN, was hit by U-19, off Rathlin by a torpedo fired at close range.

Spiess evaded a screen of seven destroyers, 11 trawlers and three sloops to finish off his stricken prey with two more torpedoes. HM Trawler *Corrie Roy*, Lieutenant Robert Baty RNR, hastened to the scene, a loud explosion having been heard, a large volume of smoke

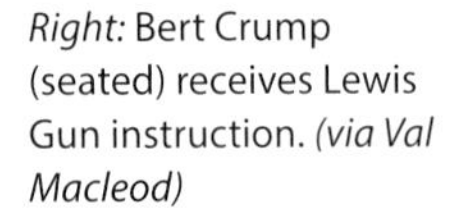

Right: Bert Crump (seated) receives Lewis Gun instruction. *(via Val Macleod)*

Far right: An airship pilot needed protection from the elements. It is believed that this photo was taken of Flt Sub Lt Bert Crump. *(via Tom Jamison)*

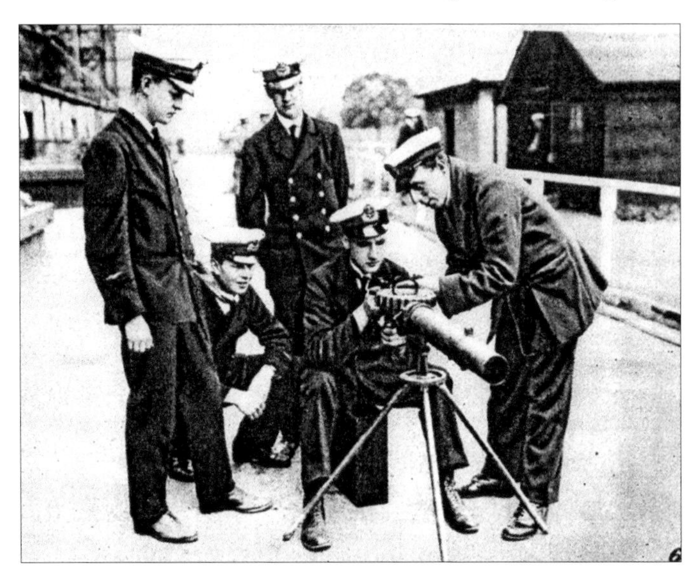

SS *Calgarian* in merchant navy service. *(Author's Collection)*

appearing around the *Calgarian*, which was steering a very erratic course. On closing a sloop signalled, 'Keep sharp lookout for submarine.' Two of the crew, 2nd Hand William Butler and 2nd Engineman William Macintosh, saw the U-boat's periscope about 150 yards abeam of *Corrie Roy*. Baty ordered that they should engage with the 7.5-inch howitzer, which fired small bombs capable of exploding 40-60 feet below the surface. Following the track of the submarine, he dropped three depth charges and then broke off to assist HM Trawler *Thomas Collard* which had the *Calgarian*'s survivors on board and was herself sinking. The 143 survivors were taken to Londonderry by HM Trawler *War Duke* out of Buncrana.

One of the lucky ones was Lieutenant Henry Kendall RNR, another incident in an eventful career, surviving two shipwrecks and being instrumental in the arrest of the notorious murderer Dr Crippen, who had attempted to flee justice with his lover, Ethel Le Neve, on Kendall's ship, SS *Montrose*, which received a description of the pair via a wireless dispatch. Chief Inspector Walter Dew of Scotland Yard, in hot pursuit, used the SS *Laurentic*'s speed to arrive in Canada before the fleeing suspect on the SS *Montrose*. Crippen was arrested, convicted of his wife's murder, and hanged.

The monthly report for Luce Bay noted that two Airedale terriers had been acquired 'for the better protection of the airship shed at night.' Engine trouble afflicted Flight Lieutenant Pullan again on 3rd March, when he had to land SSZ12 at Bentra for repairs.

The following day FSL Crump carried out an 11-hour patrol in SSZ13 which included escorting the *Princess Maud* – a duty which he was to repeat more than once over the next few months. Seventeen patrols of greater than eight hours duration were undertaken that month. The first bomb to be dropped in anger from a Luce Bay airship was on 18th March when SSZ11, flown by FSL SB Harris, attacked a suspected submarine off the Copeland Islands at the mouth of the south side of Belfast Lough. An airship communicated by radio on the evening of 18th to the naval forces assembled, again off the Copelands, the sighting of a large patch of oil, which *Zara* depth charged. Harris made another bombing attack before

Lt Pullan was a regular visitor to Bentra. *(D&N Calwell Collection)*

An SSZ Class airship, looking over the pilot's head to the engineer's station at the rear of the car. *(via Tom Jamison)*

The *Princess Maud* nears harbour escorted by an airship. *(via Ernie Cromie)*

the end of the month while escorting a north-bound convoy near the entrance of Belfast Lough. Also in March Commodore (Second Class) Charles Carpendale RN was formally appointed to *Vigorous*, the Auxiliary Patrol Depot Ship at Larne, in charge of Auxiliary Patrol Area XVII. Between 1915 and 1917 he had been Chief of Staff, Coast of Ireland and Flag Captain Queenstown. He was much younger than his predecessors being 43 years old.

By this time the North Channel Patrol consisted of six destroyers including HMS *Avon*, HMS *Express*, HMS *Osprey* and HMS *Wolf*, as well as HMS *Dove* and HMS *Thorn*. A motor launch hunting flotilla was now based in Larne, organized usually in groups of three, the senior ship being fitted with wireless and all being equipped with hydrophones, a gun and depth charges.

Other Irish bases

Londonderry had become the base for the 2nd Destroyer Flotilla and the 2nd Sloop Flotilla, 28 warships in all, which previously had been stationed at Buncrana.

It also provided facilities for docking and refitting the smaller patrol vessels. One of the more important companies carrying out this work was Craigs Engineering Works of Queen's Quay. For example, in 1918 the armed trawler *Imperial Queen* was fitted with a 7.5-inch howitzer (anti-submarine mortar) and hydrophones to add to its 12-pounder. Further round the north coast to the west a maximum of 47 smaller patrol craft, trawlers, drifters and boom defence vessels, were based at Lough Swilly, Auxiliary Patrol Area XVIII, which also

Three motor launches in the Irish Sea. *(National Archive)*

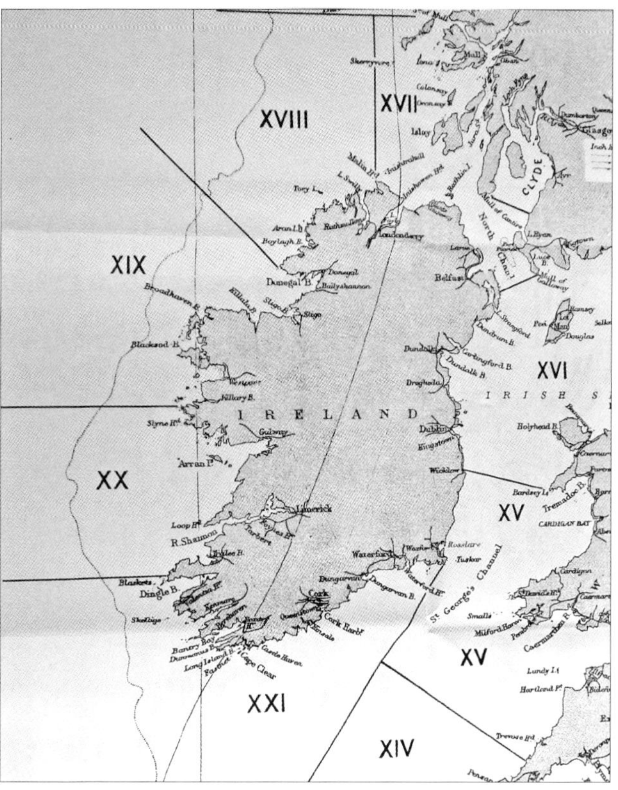

controlled the smaller base at Killybegs. The southern part of the Irish Sea fell under the remit of Auxiliary Patrol Area XVI at Kingstown (now Dún Laoghaire). Auxiliary Patrol Area XIX, based at Blacksod Bay in Co Mayo and Auxiliary Patrol Area XX, at Galway Bay in Co Galway, covered the west coast, while Auxiliary Patrol Area XXI covered the south and south-west coasts, based at Queenstown and Berehaven in Co Cork.

During 1918 another submarine depot ship, HMS *Ambrose*, was also stationed in Ireland, having arrived at Berehaven on 2nd December 1917. This became for a time a combined *Vulcan/Ambrose* Flotilla.

Above left: The submarine depot ship HMS *Ambrose*. *(Author's Collection)*

Above: The Auxiliary Patrol Areas around Ireland. *(via Anthony Kinsella)*

An unusual cargo

A most unusual escort duty took place on 29th March 1918, when an airship accompanied the *Princess Maud* bearing a unique cargo from Larne to Stranraer. The story had begun some six months earlier, in August 1917, when the War Cabinet authorised the Handley Page Company (HP) to construct three prototypes of a large strategic bomber, even bigger than its O/100 and O/400 twin-engine heavy bombers. As the HP factory at Cricklewood in North London was working to full capacity, it was decided to make use of the expert draughtsmen, carpenters and fitters of the shipbuilders, Harland and Wolff Ltd, in Belfast. To this end, the Chief Designer, GR Volkert, went there to take charge of the project at the head of a team from HP, while Frederick Handley Page himself visited the city every weekend to monitor progress He returned to London by means of the *Princess Maud* and the railway from Stranraer on Mondays with all the Ulster ham and bacon he could carry.

Harland and Wolff also already had relevant experience in aircraft manufacture, having commenced in 1916 the construction of an eventual total of nearly 1000 single engine DH6s and Avro 504s, which were shipped by water to England.

On 28th February 1918, Aldergrove in Co Antrim was selected as the site for the flying test field for the new bomber, with the result that the construction of final assembly hangars was needed. The flight sheds were not ready in time for the first prototype, so the massive HP

The Handley Page V/1500 was flown from Aldergrove as well as being sent by sea. *(Ernie Cromie Collection)*

A DH6, the first aircraft built by Harland & Wolff. *(UFTM)*

V/1500 was transported in parts to Cricklewood (the fuselage was shipped directly to London from Belfast but the 2800 square feet wings went by way of the *Princess Maud* from Larne and then by rail to Euston) where it flew for the first time, registered as B9463, on 22nd May 1918.

Bert Crump also carried out a patrol on 29th March, in SSZ20, from six miles SE of the Copelands to six miles north of Rathlin Island, lasting nine hours and 15 minutes. His Log Book notes that he escorted SS *War Expert*, a brand new tanker of 5259 tons, which was sunk by U-402, Kapitänleutnant Freiherr Siegfried von Forstner, twenty four years later during the Second World War, having been renamed SS *Empire Progress*. He then escorted SS *Cedric*, which, when she was launched in Belfast in 1902, was the largest ship in the world. On what was rather a busy day, SSZ11 dropped two bombs one mile east of Whitehead, with an armed trawler adding three depth charges in the vicinity of an 'oil streak'.

From the RNAS to the RAF

Meanwhile, on 31st March 1918 came the final patrols by personnel of the RNAS. It had been a record month with a total of 324 hours and 9863 miles flown by the four airships, escorting 10 convoys and 177 ships sailing singly, with 77 patrols being of more than eight hours in duration and three of over 10 hours. From the following morning these activities were conducted by the Royal Air Force (RAF). They were the same crews but the RNAS and the Royal Flying Corps (RFC) had merged to form the RAF. One significant change was the adoption of military ranks. Another was the closure of the bank account at Whitehead in

The *Princess Maud* with escort in 1917. *(Donnie Nelson Collection)*

the name of the RNAS on 17th April, the final authorised signature being that of Assistant Paymaster GR Renniser, who had taken over this administrative function from Flight Commander Hartford in January.

Oddly enough, the Admiralty retained responsibility for the actual airships until October 1919. Patrol and escort duties continued with two more notable landings at Bentra. On 9th April, Lieutenant Pullan had to land SSZ12 for more repairs, this time to fix a damaged elevator. On the same day Lieutenant Crump in SSZ20 was escorting a convoy near Rathlin Island when he witnessed a trawler dropping depth charges on a suspected submarine contact. On 16th April Major Irving Hartford made his final sortie as CO, flying from Whitehead in SSZ20, escorting the *Princess Maud*, having spent the night at Bentra.

Victualling under the Air Force system had some teething issues, 'Various difficulties arose with the purchasing of certain supplies of foodstuffs through the Navy and Army Canteen Board, which, combined with the fact that definite articles of field rations are not reliable, made the adoption of the 'Universal Diet Sheet' impossible. In spite of the difficulties a varied and satisfactory menu was supplied for the men.' It was also remarked upon that paperwork generally had increased, with reports now having to be made to the Air Ministry and the Admiralty.

Chapter 8

Two U-boats are sunk by Larne-based vessels

IN THE LATE AFTERNOON of 17th April 1918 the drifter *Pilot Me*, Skipper Andrew Walker RNR, off Torr Head, spotted a periscope fifty yards distant to starboard. The drifter turned towards the target and depth-charged the already diving submarine whilst zigzagging over the course the enemy appeared to be steering. She then stopped and used her hydrophone. Fifteen minutes later, UB-82, Kapitänleutnant Richard Becker, broke surface between the *Pilot Me* and another drifter, *Young Fred*, Lieutenant Thomas Kippins RNR. Met with gunfire from other drifters as well, particularly *Light*, the damaged submarine tried to dive again. It was too late; *Young Fred* dropped another couple of depth-charges over her side. This caused so terrific an explosion and such a high column of water that the other drifters believed for a moment that the *Young Fred* was gone. However wreckage shot up from the shattered submarine, such as woodwork fittings, gratings, and seamen's caps bearing the words 'Unterseeboots Flotilla'. The submarine had been so damaged that she broke surface, and remained exposed at a very steep angle until the *Young Fred* finally dispatched her. Commodore Carpendale arrived with HMS *Express* and found a division of drifters gathered around a patch of sea covered with thick oil and odd items of debris, with a heavy smell of fuel in the air. Kippins and Walker came on board to report. As well as describing the action, they told the Commodore that they had picked up from the sea, 'Two German

Below: UB-82 to the left and UB-85 to the right. *(U-Boot Archive)*

Below right: The skipper and crew of an armed drifter. *(Author's Collection)*

HMS *Thetis (Author's Collection)*

seamen's caps, clothing, bedding, internal woodwork fittings, a red flag and two German postcards.' The Admiralty considered this 'a very creditable and successful operation,' and added, 'a number of DSCs and DSMs were distributed, and the sum of £1,000 was awarded to those who had brought about this commendable achievement.' Walker and Kippins both received a DSC, and a deckhand from each of the drifters the DSM. Skipper RNR was a rank equivalent to Warrant Officer.

On 18th April SSZ12 was reported at Bentra again, this time with a gas leak. Hartford's successor, Major William Pennefather, carried out his first escort duty on 23rd April in SSZ20. St George's Day 1918 was also the date of a bold and valiant attempt to deal with the U-boat menace by other means. The plan was to block the enemy's access to the sea from its base at Bruges by blocking the canal mouth at Zeebrugge with sunken warships.

One of the blockships was the twenty-eight year old HMS *Thetis*, an *Apollo* Class cruiser, which had previously served as the depot ship at Larne Naval Base from 1916 onwards. The early patrol that morning had been flown by Lieutenant Crump in SSZ12, who was making his first visit to Bentra. A few days later, on 26th April, he had to land SSZ20 at Larne when the airship's engine developed carburettor trouble.

The UB-85, *Kempock* and *Coreopsis*

Late on the evening of 28th April HM Armed Drifter *Willowbank*, Lieutenant FS Cole RNVR, picked up a hydrophone contact off Black Head, a U-boat was seen on the surface, *Willowbank* opened fire and gave chase but was outrun by the submarine. The following day the SS *Oxonian*, while under escort from the Armed Trawlers *Lewis Reeves* and *Lewis Roatley*, was attacked by two torpedoes off Corsewall Point. Luckily they missed, as also did the depth charges dropped by the escorts. Early on the morning of 30th April the German Type UB III coastal torpedo attack submarine, UB-85, Kapitänleutnant Günther Krech, was on patrol off the County Down coast. On sighting the steam coaster *Kempock*, off Ballyferris, the U-boat surfaced and engaged the target with its 88 mm deck gun. The 52 year old, defensively-armed, iron screw steamer, which was carrying a cargo of potatoes from Belfast to Manchester, fought back with its recently mounted, 4-inch gun and a two hour battle ensued. UB-85 was in the end victorious but at high cost, as it proved impossible to submerge, so great was the damage caused by the valiant *Kempock*, which sank six miles SE of Copeland Island Light. Captain John Roberts and his crew abandoned ship and reached Donaghadee safely. Roberts received the DSC and the crew a monetary award.

As the U-boat limped north it ran into HM Armed Drifter *Coreopsis*, Lieutenant Percy S Peat, RNR, which was armed with a single six-pounder, on its first patrol out of Belfast. The drifter opened fire at point-blank range, the first and third shots exploding. Accurate shooting was difficult owing to the heavy swell. The enemy rapidly drew away, then stopped suddenly and fired a white Very light. On approaching the submarine, Peat's crew heard the Germans all shouting together, 'We will surrender'; 'We are your prisoners.' Lieutenant Peat was so surprised that he could only assume the submarine was damaged, for no attempt had been made to respond to the British attack. He passed close astern of the submarine, the Germans continuing to shout that they desired to be taken prisoners. Peat manoeuvred his vessel so as to bring the enemy into the moon's rays and again opened fire, for the U-boat's gun was plainly visible, and it was, as a precaution, necessary to put it out of action. Four or five British shells hit the target, but still no reply came from the enemy. So approaching within hailing distance, with his gun trained on the submarine, Lieutenant Peat informed the enemy that, if the submarine attempted to move, she would be fired on again.

At 4.00 am other patrol craft began to arrive on the scene, HM Armed Drifter *Valorous*, HM Trawler *Lewis Reeves* and HM Trawler *William Symonds*. Peat now sent away his boat with three hands to approach the submarine, but not to close her. He then told the Germans to jump into the sea and the boat would pick them up. The order was obeyed, and thus nine prisoners were brought on board. But in the meantime the German commanding officer and some of his men tried to remain to sink the submarine. Altogether the boat of the *Coreopsis* took three officers and twenty four men prisoners in the most unexpected circumstances.

UB-82 with UB-89 behind.
(U-Boot Archiv)

As they came on board the drifter, Lieutenant Peat searched each one carefully for papers and arms and then signalled the Senior Naval Officer at Larne. Escorted by other auxiliary craft, the little ship subsequently had the privilege of entering Larne with her captives. So a U-boat armed with a 4.1-inch gun and torpedoes, carrying a crew of thirty-six, surrendered to a small, slow surface vessel armed with only a 6-pounder gun, and a crew of twelve.

Krech explained to his captor that the submarine's crew had been suffering from the effects of excess chlorine gas being given off from the batteries. Moreover, the conning-tower had been jammed by one of the *Kempock's* shells, so when UB-85 tried to submerge in order to escape, fifteen tons of water poured in, which did little to improve the efficiency of the already defective batteries. UB-85 was sunk by gunfire nine miles east of Blackrock, Islandmagee. For the Germans, the war was over and they were landed at Larne before transfer to a prisoner of war camp. Lieutenant Peat received a DSO, his second-in-command, Skipper GE Stubbs, RNR, the DSC and the 2nd Hand, David Christie, a DSM. A sum of £1000 was divided among the crew. The news of Peat's DSO was very well-received in Larne, with the *Larne Weekly Times* commenting, 'he deservedly holds a high place as a gallant and successful officer, and his many Larne friends were intensely gratified when he was recently gazetted.' Krech never saw Germany again, as he died in hospital in England in March 1919 at the age of 33.

On 1st May HMY *Albion III* was attacked by a U-boat while returning from escort duty. The track of the torpedo was plainly seen and the helm was put hard a port, causing it to miss by a few feet. Two depth charges were dropped but the U-boat got away. A few days later, on 5th, the SS *Greenisland* reported having rammed a supposed U-boat off Bengore Head, near Bushmills on the North Antrim Coast. It is believed that this was UB-119, Oberleutnant zur See Walter Kolbe, and that it sank with all hands.

A message from the same ship, three days earlier, had reported sighting a U-boat on the surface off Rathlin. Nearby on the same day the SS *Wheatear* sighted a submarine and fired

The gallant SS *Greenisland*. *(Courtesy Graham Walton)*

two shots at her. The U-boat returned fire and a running battle ensued as *Wheatear* ran for Portrush, making smoke as a defensive shield. Several shells landed on the coast near the village of Portballintrae. Commodore Larne ordered HMS *Express*, HMS *Osprey*, HMY *Zara* and three trawlers to patrol between Rathlin and Inishowen but no further contact was made.

After hostilities had ceased it was possible to tell many stories which had been restricted by the wartime regulations. Under the heading, 'Four Times Torpedoed – Yet Able To Reach Port – Belfast Crew's Ordeal', an article appeared in the *Belfast Weekly Telegraph* on 30th November 1918:

> 'Now that the restrictions of censorship have been largely removed, Belfast citizens may learn of a thrilling adventure that befell Messrs G Heyn & Sons steamer *Rathlin Head*, some time ago when she was attacked by an enemy submarine. The story of that grim duel near the Irish coast reflects the highest credit both upon officers and crew and adds yet another laurel to those held by the men of the mercantile marine. In connection with this episode it may be mentioned that Captain WJ Campbell of the *Rathlin Head* was later decorated with the Distinguished Service Cross. It was on a Saturday afternoon, and the ship was ploughing her way through a friendly sea when she first encountered an ocean pest, whose "visiting card", as usual, consisted of a couple of torpedoes dispatched on an errand of death to an unsuspecting steamer. These missiles struck the *Rathlin Head* on the starboard side close to the No 1 hold which was immediately filled to sea level, the wounded vessel taking a forward plunge that brought her propellers almost out of the water.'

The ship which had been built by the Workman, Clark and Co shipyard in Belfast, had been on passage from Milford Haven to New Orleans in ballast. The submarine on 25th May 1918 was Kapitänleutnant Leo Hillebrand's U-46. The article continued:

> 'At 5.30 pm, about an hour and a half after the first attack those aboard the *Rathlin Head* sighted the submarine on the starboard quarter and fire was opened on her with a 4.7 inch gun but the submarine was well out of range and the shells fell short. In the meantime SOS messages had been sent out and presently a reply was picked up from the radio station at Crookhaven [in West Cork] stating that the tug *Flying Spray* was on its way to render assistance. At midnight came the cheerful news by wireless that a destroyer had picked up the SOS and was rapidly approaching. The Chief Officer, Mr Dudley F Moore, vividly describes the incidents that followed, 'All went well during the night, all hands remaining on deck all night and we were in occasional communication with Crosshaven and other radio stations. At 7.25 on the Sunday morning I was standing alongside the captain on the port side of the bridge when we saw two torpedoes coming directly for the ship, and though the helm was immediately put hard over, they struck her about the stokehold, immediately

USS *McCall (Library of Congress)*

> flooding the main bunker, stokehold and engine-room to sea level, throwing a dense cloud of water out of the funnel and a volume of water onto the bridge deck. This explosion flooded the engine-room and killed the three firemen who were on watch – Issac Cinnamond, Thomas Milliken and William James Rawe. The boatswain was standing at the port howitzer when he saw the torpedoes coming for the ship and just before they struck her, he fired at the spot the torpedoes came from about 300 yards off, and immediately after the shell struck the water a dense volume of black smoke or oil rose in the air and the shell did not ricochet. Just after the torpedoes hit the ship she took a heavy list to port and the captain ordered everyone to the boats.'

After about a further twenty minutes the crew of the stricken ship saw smoke on the horizon, which proved to be the destroyer USS *McCall*. At some considerable risk to itself the *McCall*, Lieutenant Commander Fred T Berry, hove to and picked up those of the *Rathlin Head*'s crew who had taken to the lifeboats.

As the *Rathlin Head* had righted itself in the water, Captain Campbell and his senior officers, decided after inspecting the damage, that a tow was a possibility. In the meantime further torpedoes were fired at the *McCall* and at the tug *Cartmel*, which had also arrived on the scene. Campbell ordered his gunners to fire the howitzers again while the towlines were made fast to the two tugs. This was accomplished and a further line was passed to the Armed Trawler *George Andrew* to assist with steering. By midnight *Rathlin Head* had been brought safely to shore at Berehaven.

Irish Sea Hunting Flotilla

A new factor had entered the equation as Admiral Bayly described:

> 'In 1918 Captain Gordon Campbell instituted at his own suggestion the Irish Sea Hunting Flotilla. We knew that submarines were coming through the North Channel

The scout cruiser HMS *Patrol*. *(NMRN)*

with the tide, so that by stopping their engines altogether or greatly reducing speed they could slip past without being detected by our hydrophones. Captain Gordon Campbell's flotilla, therefore, was of great assistance to me, relieving the Queenstown force of a considerable area to cover. No matter what he was called on to do, Captain Gordon Campbell was always willing and ready to do it, fearless of responsibility, and he commanded my admiration. I gave him a completely free hand with the hunting flotilla. It might have been brought into operation earlier had the units been available. It consisted of the light cruiser HMS *Patrol*, of which Campbell was the Captain, a flotilla of old destroyers, a yacht, a drifter, twenty motor-launches, four airships, and a squadron of aeroplanes, operating from Kingstown and Holyhead. He also had the help (in due course) of some of the US sub-chasers. The airships watched over the sea, and the aeroplanes near the coast. Quite apart from their speed, aircraft proved very useful in detecting submarines because they could see so much farther than ships. From the air they could also detect submarines even when submerged, provided the water was clear. On the other hand, submerged submarines could not see aircraft, even when they had their periscopes above the water. In addition to the material help they gave, the aircraft undoubtedly had a great moral effect on the submarines' crews owing to the fact that the airmen could see without being seen.'

An oil slick photographed at 400 feet an airship after dropping bombs on a suspected U-boat sighting. *(National Archive)*

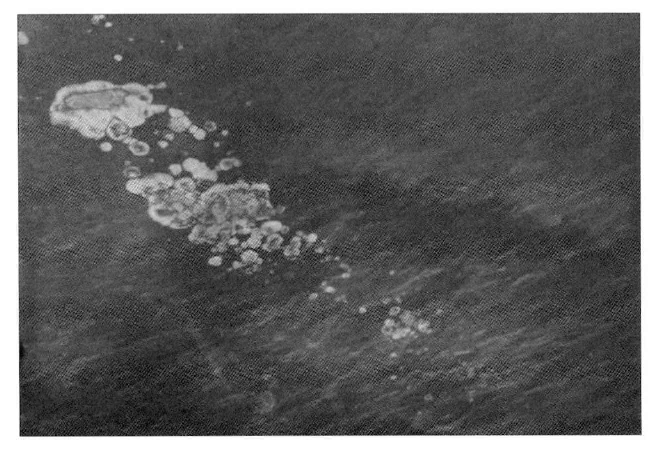

On 28th May Campbell wrote a report to Bayly about 'a smart piece of work' which, in his opinion, was a fine example of air/sea co-operation within his command. The destroyer HMS *Albacore*, Lieutenant-

Above left: HMS *Earnest* *(Author's Collection)*

Above: The C Class destroyer, HMS *Kestrel.* *(NMRN)*

in-Command LW Newbery-Boschetti, was working with several MLs and two airships, SSZ35 and SSZ51, dropping bombs and depth charges on an oil patch, where is was suspected that a U-boat lurked. SSZ35, Lieutenant R McMurray, suffered engine failure and was towed back to Holyhead by ML354, Lieutenant AE Lorimer, the tow taking some two and a half hours.

It should be noted, however, that Commodore John Denison (who had replaced Herbert Chatterton at Kingstown in mid-1917) had, in March 1918, requested the allocation of faster vessels such as destroyers to supplement his armed trawlers, drifters, motor launches and yachts in the light of increased U-boat activity in the Irish Sea, with the result that the destroyers HMS *Kestrel*, HMS *Zephyr*, HMS *Seal* and HMS *Earnest* were dispatched to Kingstown on 16th March, accompanied by the depot ship HMS *Vulcan*.

Meanwhile, the previous month's record for the RNAS airships at Luce Bay and Bentra was broken by the RAF in its first month. The four airships carried out 132 patrols, flying 688 hours and 20,525 miles, with SSZ11 alone topping 200 patrols and 6000 hours. Definite progress was being made, as Chatterton notes,

> 'Now pass on a few weeks to the spring. What is the condition of affairs? The convoys were well-protected, the U-boats were gradually having their life squeezed out of them, and the shipping losses now were very different from what they had

> been. Hydrophones, deep-laid mines, ever-vigilant destroyers, P-boats, trawlers, drifters, yachts, MLs, disguised Q-ships, depth charges, howitzers, many more guns, and better defensive methods for the Mercantile Marine ships: all these items were beginning to tell heavily on the enemy's U-boats. He realized it; he knew to his cost that month after month his craft went forth and never came back, so that the Flanders submarines were known among their own officers as the "Drowning Flotilla".'

Patrol work

On 5th May, SSZ12 was returning from a Northern Patrol when very high winds were encountered. Running full out the airship was unable to make any headway, so an emergency landing was made at Whitehead in the gathering darkness. As only a relatively small landing party was available the envelope had to be ripped open to effect a swift deflation. The partially dismantled airship was returned by sea from Larne to Luce Bay. It was flying again three days later. SSZ12 called again at Bentra in more controlled circumstances on 14th May. SSZ11 made a forced landing on the Isle of Arran on 20th May due to engine failure, 'the trail rope and drogue were dropped and two green lights fired. The drogue brought the ship up but shortly afterwards it was carried away and the drifter which was coming up was unable to get hold of the trail rope. The ship made a free balloon-landing, the only damage being a broken rudder and propeller.'

An inconclusive report from Buncrana to Queenstown on 23rd May noted that a submarine had been sighted by an airship, bombs had been dropped, as well as three depth charges from a surface vessel, and a large quantity of oil had come to the surface. Lieutenant Crump in SSZ20 had another taste of naval pyrotechnics on 27th May when he sighted bubbles of oil while escorting the *Princess Maud*. A trawler and a drifter dropped eight depth charges. Not to be left out, he unloaded a couple of 65 lb bombs himself. On 30th May he noted that the inbound convoy, which he overflew for eight hours at a height of 1000 feet, was carrying 35,000 troops from the USA. Gibson and Prendergast state that, 'through the North Channel in 1918 passed one continual stream of American troopships, while the enemy submarines made great efforts to harry them. The Germans complained particularly of the numerous British airships and seaplanes; they did not fear bombs being dropped on them, but they did dislike having their movements watched and reported. Over 1,500,000 US troops were conveyed to Europe in the summer of 1918.'

On the same night as Crump flew over the convoy 12 fishing boats from Kilkeel, Annalong and Portavogie in Co Down were a dozen miles off the coast with their nets out, when a U-boat came out of the mist towards them. It was UB-64, Kapitänleutnant Otto von Schrader, from which the order was heard, 'Lay to and heave no line.' One by one the little drifters were boarded and small charges laid. The crews were treated with courtesy and plied with gin and cigarettes. Two of the boats, *Moss Rose* and *Leonora*, were spared to allow the crews to reach the shore at Kilkeel and Ardglass.

In June 1918 the *Princess Maud* was fitted with self-protection equipment of a rudimentary kind, machine-guns and smoke generating apparatus. A gun crew was also supplied. The Admiralty further instructed the owners that all crossings during the hours of darkness should be made at full speed with no lights showing. In the event the month brought something of a respite when compared to the level of activity during the spring. This was not to last, as on 28th June the RN submarine D-6 of the North of Ireland Patrol was torpedoed and sunk by UB-73, Kapitänleutnant Karl Neureuther.

July began quite excitingly for Lieutenant Crump. On 4th of the month he was escorting the *Princess Maud* from a height of 3800 feet in SSZ20 when the engine stopped. There was a problem with the crank handle, flywheel and crankshaft. He re-started the engine by kicking the propeller with his foot.

A chart has survived showing the 19 patrols made from Luce Bay during the week ending 13th July 1918. Six round trips were made to Bentra, while patrols along the Ulster coast ranged as far north as Rathlin Island and as far south as St John's Point on Dundrum Bay. 110 hours and 26 minutes were spent in the air by SSZs 11, 13 and 20, covering 2467 miles.

Lieutenant Crump had a further mishap on 15th July, which his log book records laconically

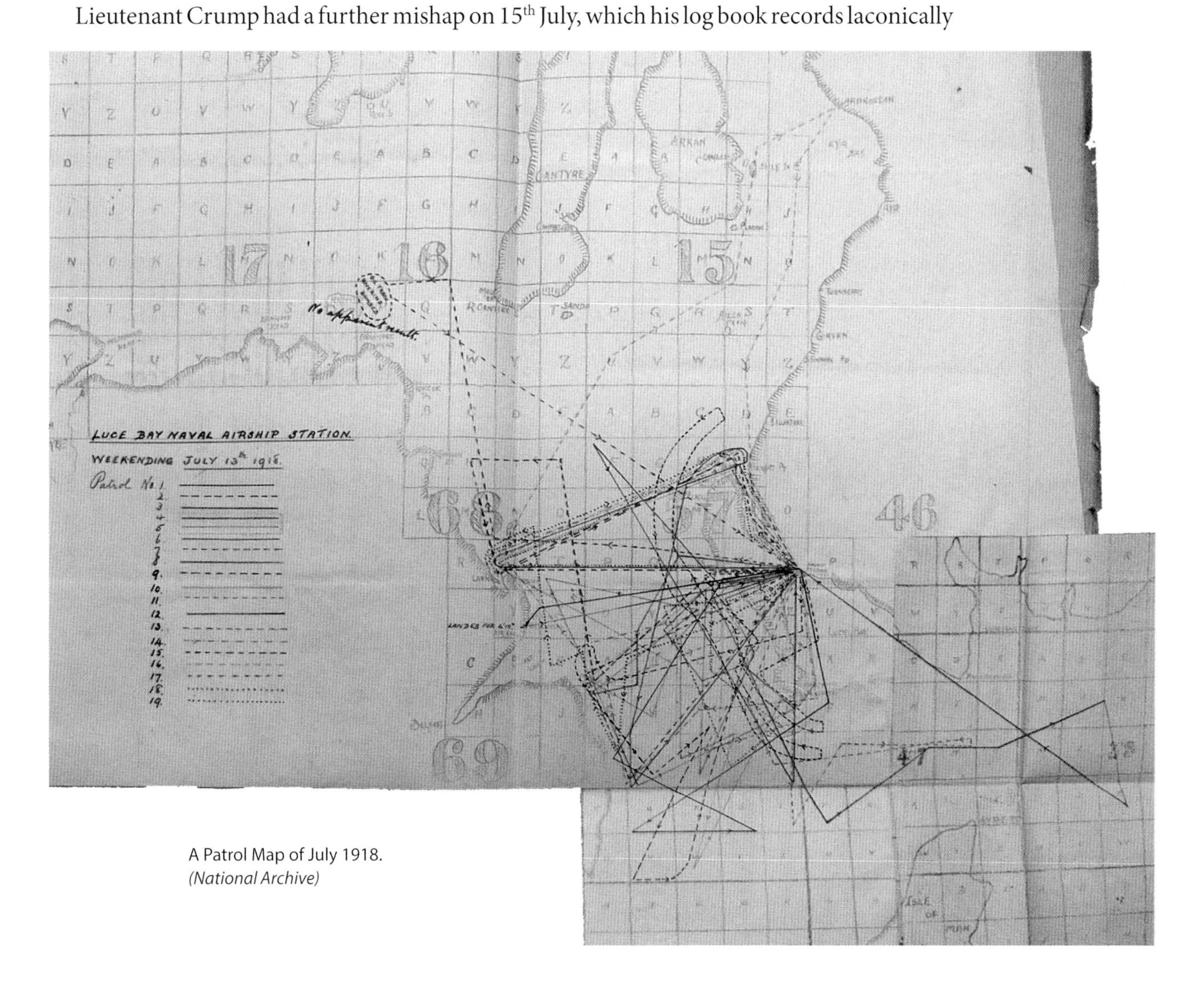

A Patrol Map of July 1918.
(National Archive)

Above: A U-boat of the same class as U-54 and U-55. *(Author's Collection)*

Above right: A sloop in the Irish Sea. *(National Archive)*

with the single word 'Wrecked.' He was piloting SSZ12 at the end of a north-western patrol. Having escorted the *Princess Maud* into harbour, he collided with the flagstaff on the West Pier at Stranraer, as he altered course to return to Luce Bay, causing damage to both the car and envelope – which caused quite a sensation among the large crowd of people gathered there. He suffered no worse then bruised pride, both he and SSZ12 were back in the air soon.

Seven U-boats concentrated in the approaches to the North Channel in July. Several important ships were sunk, including the sloop HMS *Anchusa* by U-54, Kapitänleutnant Helmuth von Ruckteschell, and the Cunarder, RMS *Carpathia*, of *Titanic* fame, by U-55, Kapitänleutnant Wilhelm Werner.

After the war both von Ruckteschell and Werner were placed on a blacklist of those suspected by the Allies of war crimes, such as continuing to fire on merchant vessels after they had surrendered. On 19th July the huge, new, White Star liner, SS *Justicia*, the U-boats' second largest victim, (constructed by Harland & Wolff in Belfast, the wreck of which today lies 28 miles NW of Malin Head) had left Belfast for America and was torpedoed by Schrader's UB-64, 20 miles W by N from Skerryvore. After being hit by four torpedoes from

The British troop transport *Justicia* sinking on 19 July 1918 20 miles from Skerryvore with a loss of 10 lives after being torpedoed. (NHHC)

M Class destroyer in heavy seas. *(Author's Collection)*

this submarine and two more the following day from UB-124, Kapitänleutnant Hans Oscar Wutsdorff, she sank on 20th despite an attempt to tow her to Lough Swilly. Both submarines got away even though a huge number of depth charges were expended by up to 40 patrol craft that had arrived at the scene. UB-124 escaped by diving right under the sinking liner.

But not for long, for on 20th July the submarine was severely damaged by depth-charge attack off the North Coast of Donegal by HMS *Marne*, Commander George Bibby Hartford, HMS *Milbrook* and HMS *Pigeon*, Lieutenant-in-Command Leslie Lonsdale-Cooper, all from the 2nd Destroyer Flotilla. She surfaced and was scuttled by her crew.

A typical UB III Class coastal submarine. *(Author's Collection)*

Chapter 9

Kite Balloon Bases and Fixed-wing Aircraft

Aerial activity was by no means confined to the airships operating from Luce Bay and Bentra. The kite balloon or drachen was invented by a German Army officer, Major August von Parseval in 1894. A spherical tethered balloon has the tendency to rotate about its vertical axis thus making observation difficult and also possibly affecting the observers in the basket with nausea. The drachen overcame this effect by elongating and streamlining the hydrogen-filled balloon envelope and adding a vertical lobe to the rear, containing air collected by a forward facing scoop. The nose was thereby kept head on to the wind and the balloon was much more stable in the same fashion as a weathercock.

Towards the end of 1917 the Admiralty commenced construction of two kite balloon stations, one at Berehaven and the other at Rathmullan on Lough Swilly. These two locations were selected as both were to be the assembly points for convoys to and from the United States. The kite balloons could thus easily be taken on board the warships before the convoy sailed out to sea. They could be used in practically any weather and the plan was that at least

A British Kite Balloon being walked through the countryside near Buncrana. *(Library of Congress)*

two warships in every convoy should be equipped with them, which would be towed at an altitude of between 500 and 800 feet. The task of these manned kite balloons was to act as lookout and give an increased field of vision over the ocean for convoy escort ships of some 28 miles, with a communication link to the parent vessel. The towing destroyer or sloop carried cylinders with which to top up hydrogen gas lost by the balloon, and thus maintain operations.

During high winds, serious loss of hydrogen was experienced whereas on a calm day there was hardly any loss and consequently little need of re-inflating the balloon. The practice was to carry four cylinders for each day of operation, which on average was found to be adequate. Kite balloons were also used by trawlers to facilitate their work of finding and sweeping mine fields.

Berehaven, on a site of 67 acres two miles to the east of Castletownbere at Mill Cove, was built as a six balloon station and handed over to the US Navy in April 1918. Initial training for the American balloonists was over land nearby with observation balloons towed by large touring cars. Experimental work was carried out firstly with the sloop HMS *Flying Fox*, Acting Lieutenant Commander Andrew Mott, in July.

One trial ended in tragedy with the death of Ensign Charles E Reed on 13th August, who was thrown from the basket of a kite

Above: A Royal Navy Kite Balloon towed by a RNR Armed Drifter. *(Library of Congress)*

Ariel view of USNAS Berehaven. *(NHHC)*

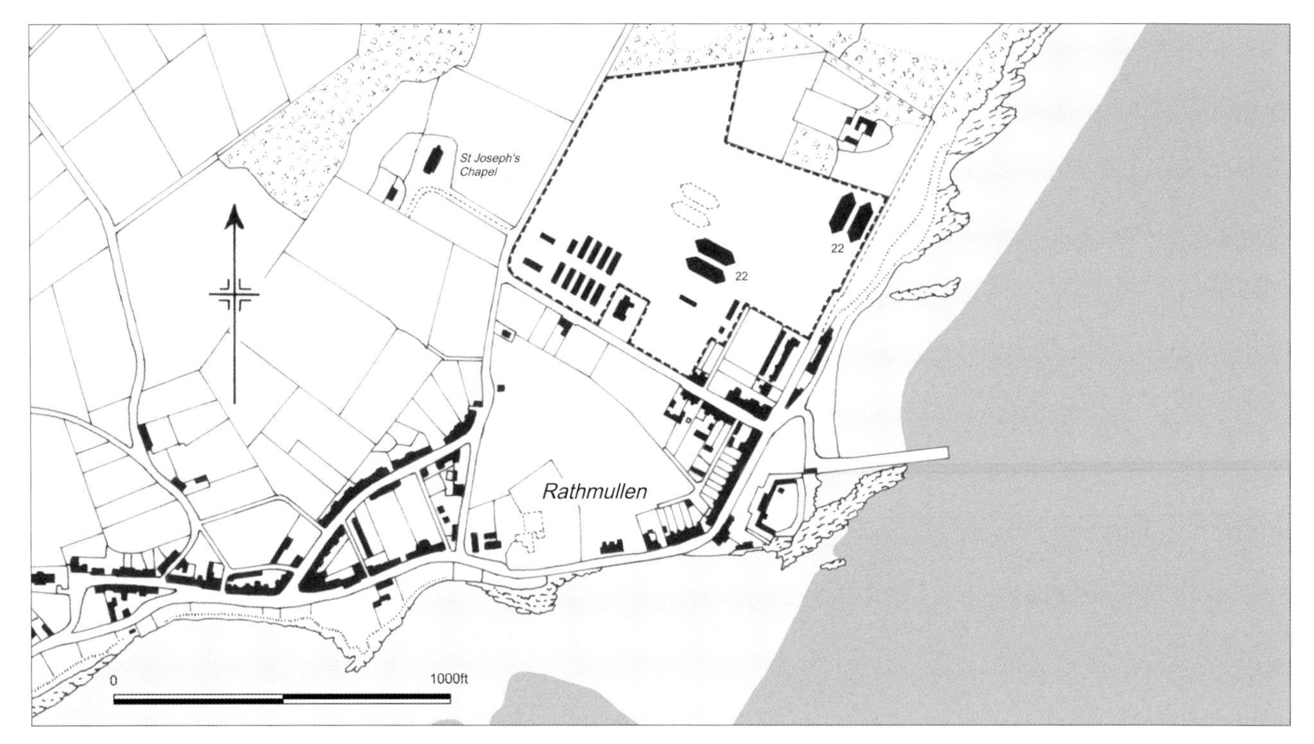

A plan of Rathmullen Kite Balloon Station in the First World War. *(Cross & Cockade)*

balloon being towed by the destroyer, HMS *Springbok*, Lieutenant Commander Cosmo Graham. Following service with the three US battleships, described later in the narrative, a skeleton crew manned the site until 14th February 1919, when it was returned to the RAF as No 17 Balloon Base. In August the RAF placed it under Care and Maintenance until July 1920.

Rathmullan was built as a four balloon station comprising four balloon sheds, each 100' 6" x 36' 0", with quarters for 125 men (14 officers being housed in the Priory) and, similarily to Berehaven, was provided with workshops, garage, oil and petrol stores, a guard hut, power house, lecture room, silicon plant, ablution hut, jetty and other subsidiary buildings on a site of 14 acres. A detachment of US personnel was also based at Rathmullan. In 1918 there was an establishment of 18 RAF officers, under the command of Rear Admiral, Buncrana. No 13 Balloon Base at Rathmullan was disbanded on 30th March 1919, with the site being also placed on a Care and Maintenance basis until July 1920.

Fixed-wing aircraft

Two RAF squadrons deployed to Ireland in May 1918. Their purpose was given as 'anti-submarine' duties. In fact the Lord Lieutenant, Field Marshal Viscount French, planned to use aircraft based in 'strongly entrenched Air Camps' for 'air patrolling with bombs and machine guns' which 'ought to put the fear of God into those playful Sinn Féiners'. The political situation was highly volatile owing to a proposal in the spring of 1918, in the light of the major German offensive on the Western Front, to introduce military conscription in Ireland and by the arrest and imprisonment of 73 leading Sinn Féin activists.

No 105 Squadron, commanded by Major Douglas Joy was sent to Co Tyrone, in mid-

Ulster, 70 miles to the west of Belfast and about as far as it is possible to get from the sea. Flights were sent to Castlebar in Co Mayo and to Oranmore in Co Galway, which at least were on the coast. No 106 Squadron, commanded by Major EAB Rice, was sent to Fermoy, some 19 miles to the north of Cork City as the crow flies and 200 miles south of Omagh.

Detached flights were sent to Athlone in Co Westmeath and Birr in King's County (now Co Offaly), both of which are nowhere near the sea. Both squadrons were equipped with RE8s, the standard Corps Reconnaissance type on the Western Front.

A few months later Captain Howard Pixton travelled throughout Ireland, across 24 counties, inspecting 65 potential landing grounds and noting carefully the nearest military or police barracks. The effect of any of the above on the war against the U-boats was negligible.

The airships were soon to be reinforced at Luce Bay by the addition of fixed-wing aircraft in the shape of A and B Flights of No 255 Squadron, which were equipped with de Havilland DH6s and soon became redesignated as No 258 Squadron. They were joined by an additional flight of the same type from No 244 Squadron. The first patrol was on 13th July.

These obsolete training aircraft, which were known not altogether affectionately as the 'Clutching Hand' or due to its slab-sided appearance, the 'Orange Box', were used to patrol the waters up to ten miles offshore. They could not compare in endurance to the airships and were only a little faster but they did add extra eyes in

Top: Major DG Joy, OC No 105 Squadron at the 'At Home' Day in Omagh. *(via Dr Haldane Mitchell)*

Above: Captain Howard Pixton *(Author's Collection)*

Strathroy Airfield, Omagh.

The DH6 otherwise known as the 'Clutching Hand' or 'Orange Box'. *(JM Bruce/GS Leslie Collection)*

the sky. As mentioned earlier Harland and Wolff had constructed many DH6s and at least one of these, C4430, flew from Luce Bay. It is not known if any of these aircraft landed at Bentra but some aeroplanes undoubtedly did. The airship station had become known locally as 'Whitehead Aerodrome' and a photograph exists of an aircraft on the ground there, taken by Jack Semple, a keen amateur photographer and nephew of James Long. It is of a bombing type flown by the RFC – the FE2b. Its appearance at Bentra presents something of a mystery. The most likely explanation is that it was manufactured by the Glasgow engineering firm, G&J Weir and visited Whitehead on an unofficial test flight – perhaps to bring back some good Ulster produce. Jack Semple also took a photograph of a BE2e, which was used by the RNAS as a trainer but this is clearly at Luce Bay. However, was Jack Semple flown over there to take his photograph?

SSZ20 visited Bentra on 23th July; the next day Major Pennefather crashed a DH6, suffering a broken jaw and nose and a dislocated thigh. He was replaced by the senior flying officer, Captain Stafford B Harris, pending the arrival of a new CO, Major EL Johnston, a qualified Master Mariner. Another squadron which was equipped with the DH6, No 272, was based at Macrihanish on the Mull of Kintyre. On 19th August Lieutenant Crump recorded that he landed SSZ20 at Bentra to replenish his ballast and on 14th September he searched for a sinking ship near the Gobbins Buoy off Islandmagee, probably the SS *Neotsfield*, which had been torpedoed by UB-64. Lieutenant E Hall of No 244 Squadron reported crash landing his DH6 F3350 north-east of the Skerries. He was retrieved from the water by the crew of the Armed Drifter *Golden Dawn*. Meanwhile, on 30th August SSZ13, crewed by Lieutenant NH Astley RAF and 1st AM Langley, was destroyed, thankfully without any casualties, after engine seizure resulting in a forced landing at Castle Head Point on the Solway Firth.

The weather in September turned unpleasant again but ships were still being sunk and patrolling continued. SSZ20 had to make a forced landing at Machrihanish on 21st September. The DH6s normally operated in pairs to provide two hour standing patrols up to 10 miles off the coast. On 22nd September 1918 the OC, Major Lister, wrote a progress report

Two views of the FE2b which landed at Bentra. It would appear that one of the RNAS ground crew is posing in the rear cockpit. *(D&N Calwell Collection)*

noting that he had recently visited the Senior Naval Officer, Larne, to arrange three standard patrol areas. The first combined patrol of two airships, SSZ11 and SSZ12 and four DH6s of 258 Squadron, C9410, C9416, C9442, and C5520, took place on the last day of the month.

Chapter 10

US Navy Aviation in Ireland, 1918

There was also an overseas air arm which made a contribution to the maritime war waged from Ulster in the Great War. On 5th January 1918 work began on the construction of Lough Foyle Naval Air Station (NAS) for the United States Navy, using local civilian labourers. It was situated at Aught Point (Lepers Point) in the townland of Ture, four miles north of Muff in Co Donegal on the western shore of Lough Foyle and nine miles from Londonderry, the nearest railhead.

A slipway and three hangars were built. The first seven enlisted men, Carpenter's Mates, commanded by Gunner HF Mosely, out of a complement of 50 officers and 450 men, reached the site on 26th February. A truck was driven 425 miles from Queenstown to help with moving the stores, supplemented by two Packard and two Tyler trucks shipped over to Dublin from the USA. A Cadillac touring car was delivered in May and on 20th of that month the US flag was hoisted. It was noted that the first baseball game was played on the Parade Ground on 30th May.

The first temporary CO was a distinguished seaman officer, Commander HD Cooke. The local workers went on strike in June, demanding higher wages, so the naval personnel took

Sailors from the US Navy in Londonderry in 1918. *(NHHC)*

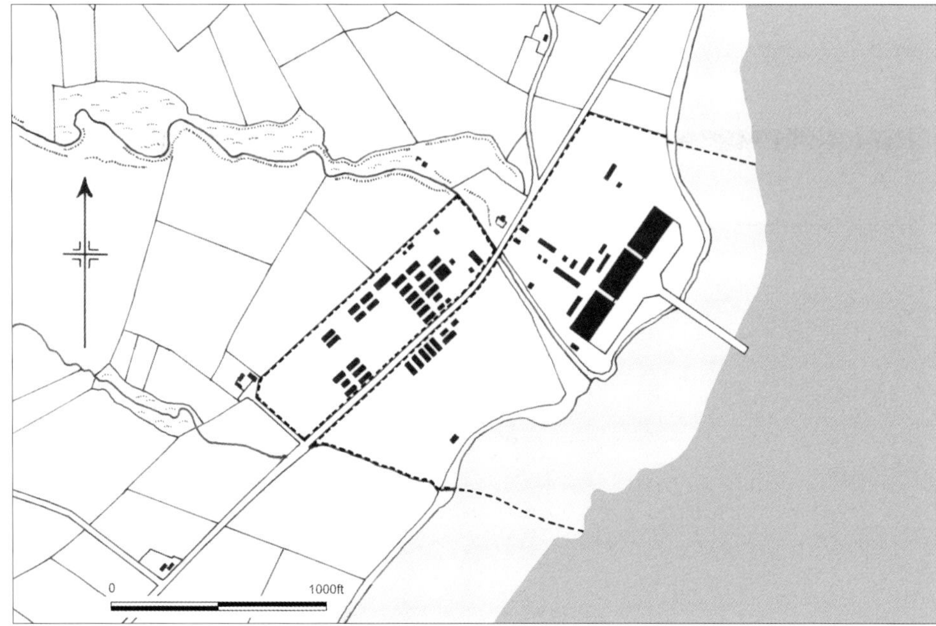

A plan of Lough Foyle NAS. *(Cross & Cockade)*

A baseball match in progress at Lough Foyle NAS. *(NHHC)*

over. The base was formally commissioned on 1st July and the men were reviewed by Rear Admiral Miller. Independence Day was celebrated on 4th July with sports and a dance to which civilian friends from Londonderry were invited, the YMCA being filled to its capacity. The 'Y' was well-appointed – with two pianos, a pool table, 75 musical instruments, three phonographs with 30 records, one motion picture machine, 180,000 sheets of writing paper and 90,000 envelopes. It also provided educational facilities on a wide range of subjects. A new CO took over on 7th July, Lieutenant CT Hull.

One notable planning error would inhibit operations. The launching ramp was constructed so that it stood above the water at low tide, preventing crews from launching or retrieving aircraft at certain times of the day. Five Curtiss H.16 'Large America' flying boats were shipped from Queenstown to Derry and trucked to the NAS, commencing 22nd July. The tardy delivery of propellers, starting cranks and radio equipment delayed the start of patrols by up to a month. The H.16 had an upper wingspan of 95 feet and a hull length of 46

Personnel at Lough Foyle NAS in the summer of 1918. *(US Navy)*

Above: An H.16 inside a hangar at Lough Foyle NAS. *(NHHC US Navy)*

Above right: Edward McKitterick. *(Author's Collection)*

feet. It was powered by two 400 hp Liberty 12 engines and was armed with four Lewis guns and four 230 lb bombs. It carried a crew of five, First Pilot, Second Pilot, Observer, Engineer and a Radioman/Electrician. It had a maximum speed of 95 mph.

The five aircraft and their assigned codes, were as follows: A-1059/LF-1; A-1031/LF-2; A-819/ LF-3; A-1032/LF-4; A-779/LF-5. The CO at Lough Foyle from 8th August was Lieutenant Edward Hyslop 'Mac' McKitterick USN, who had been born in Sissiton Indian Agency in South Dakota. He was a graduate of the Navy Academy Class of 1912 and who had set up the Aviation Ground School at the Massachusetts Institute of Technology (MIT) in 1917. He was also the 39th pilot to qualify with the USN.

Patrol lines were planned, starting abeam the lighthouse at Dungaree Head at the mouth of Lough Foyle, then west over Inishtrahull Sound to Tory Island or alternatively north-east to Oversay Island, just off Islay. Another went east to the Mull of Kintyre via Altacarry Head on Rathlin Island. The waypoints were rearranged to provide a total of six routes varying in length from 100 to 150 miles. It was noted that 'additional patrols will be added as found desirable' and 'aircraft will also be employed on escorting convoys, and when sufficient machines are available they will work in relays.'

A pigeon loft was constructed to hold 72 birds, with a compartment for two men. The birds had arrived on 15th June and were given their first flight twelve days later. In a few weeks time, the birds had become accustomed to their surroundings and were ready for training. They were trained up to a distance of 70 miles on coastlines from each side of the Lough, after which they were taken up in the flying boats for further training. After each bird had found its way home in good time from several trips, it was ready for regular patrol work. When on patrol the crews kept in direct contact with the Intelligence Officer back at base by means of radio and pigeons. The radio was fairly efficient, seaplanes receiving within 40 miles radius and the Station receiving from much further afield.

It is possible that one of the young officers at Lough Foyle was taken up for a flight by the

RAF as Captain George Bowen of No 105 Squadron, which was based at Omagh, noted in his log book that on 24th July he took a US Navy Ensign for a 'joyride'.

The US Meteorological Station at Malin Head, 35 miles to the north, the most northerly point in Ireland, was a sub-station of Lough Foyle NAS. Construction began there in May. It sent in weather reports every six hours to the NAS, which also received weather reports from the RN base at Buncrana. The relationship between the NAS and the Rear Admiral in command at Buncrana was defined thus:

Peerless, a USN homing pigeon of WW1. *(NHHC)*

> 'When aircraft are available for patrol the Rear Admiral is to be informed and whether it is desired that any particular patrols should be carried out. No seaplanes are to be sent out without reference to the Rear Admiral, and the departure and return of seaplanes is to be reported to the Rear Admiral. Test and Exercising flights may however be carried out inside Lough Foyle without reference to him. The decision as to the weather being suitable for aircraft to be sent out rests with the Officer Commanding Air Station. The Officer attached to the base at Buncrana will keep the CO Air Station informed of all information of value to air matters. CO of Air Station is responsible that pilots know the position of any of our submarines that they may sight and have the recognition signals. Communications with aircraft to be as laid down in the Admiralty Orders, many vessels have, however, not yet been supplied with Aldis Lamps.'

Buncrana and Lough Swilly received the visit of a large warship on 9th August when the old, pre-dreadnought battleship HMS *Implacable*, Commander K Toms, depot ship of the Northern Patrol, steamed in.

The pre-Dreadnought battleship HMS *Implacable. (NMRN)*

The destroyers, sloops and smaller vessels based at Londonderry and Buncrana engaged in much vital convoy work. On a somewhat smaller scale Motor Launch No 275 was sent from Larne on 11th August, 'for duties with Lough Foyle NAS.'

On 22nd August, the first test flight was made by LF-1, with Ensign A Parker Teulon, United States Naval Reserve Force (USNRF) as 1st Pilot, who was from Dorchester, Massachussets, and Lieutenant McKitterick as 2nd pilot, 45 minutes around the Lough.

Above: A flying boat is manoeuvred into the waters of Lough Foyle. *(NHHC)*

Above right: A good impression of the size of Lough Foyle NAS may be gained from this image. *(NHHC)*

The weather at the Station was seldom ideal for flying. High hills and mountains around Lough Foyle created turbulence, with strong winds, hail, rain and low hanging cloud adding to the difficulties. Further training flights took place before the end of the month. Sadly, on 31st August, Carpenter's Mate 1st Class Edward Joseph Reilly USNRF succumbed to pneumonia. On a happier note, on the same day the Station held a 'Sports Day and Fun Fete', followed by dinner and a musical play, 'Shove Off.'

The first operational flight by the USN in Ireland followed on 3rd September. LF-1, which was carrying bombs for the first time, was airborne for four hours and flew from Dungaree Head to Tory Island, the Mull of Kintyre and back. A further, slightly shorter patrol was undertaken the next day. Ensign Teulon flew LF-2 on patrol on 8th September. On 9th September an inbound American convoy, HS-53, was escorted in by LF-2, with 1st Pilot Ensign George Montgomery, who had been born in Whapeton, North Dakota. After the aircraft had left the convoy it was forced to land with engine trouble and was towed in by ML 275, which was sent in response to pigeon messages. LF-1, Ensign Teulon, then picked up the same convoy and escorted it through to the Mull of Kintyre. On the following day LF-1 landed on the water off the Giant's Causeway to effect repairs to the radio generator which had become loose. In taking

Programme

Sports Day
and
:: Fun Fete ::

U. S. Naval
Air Station,
Lough Foyle

:: Saturday, ::
Aug. 31, 1918.

Sports Day Programme for Lough Foyle NAS. *(via Ronnie Vogt)*

Right: Spectators at the Sports Day on 4th July 1918. *(NHHC)*

Far right: A senior rating with a megaphone at the Sports Day. *(NHHC)*

off again the starboard propeller was broken by the heavy sea and the hull badly damaged. Portrush Coastguard reported by signal, 'Disabled seaplane being towed. Commodore Larne informed.' The flying boat was towed into Portrush Harbour, and later back to Ture, where it was decided that it was damaged beyond repair and was therefore cannibalised for spare parts. Bad weather restricted operations over the next few days to a single training flight.

Curtiss H.16 LF2 on the slipway at Lough Foyle NAS. *(NHHC)*

The signal logs maintained by the Intelligence Officer, Ensign George Zabriskie, from New York, show that the Station was advised of convoys passing through, giving details of their direction, position, speed and estimated time of arrival. Escort duties were ordered 'as far E or W as practicable, if weather permits'. Other regular signals summarised U-boat sightings and gave detailed weather reports, forecasts and warnings of depressions from as far afield as the east coast of the USA. Intelligence information was displayed on wall charts in the Intelligence Office, so that pilots could keep themselves fully briefed. Fixes of U-boats were plotted on another chart; another showed the position of convoys and a third the general war situation. A blueprint chart was issued to the crew before an operational flight on which they would record timed positions, bombs dropped, convoys sighted etc. Every morning at 07.30 the duty telephone orderly was required to ring US Air Operations at Queenstown and advise of the number of aircraft in commission, confirm that patrols would be carried out as ordered by the Admiralty at Buncrana and give the Buncrana weather report.

The Station lost another member of personnel on 18th September, when Boilermaker William Edward Kelly USNRF drowned. The report made on this accident is poignant, 'While attempting to secure a hat that had blown into the water, the following man fell into the water and was carried by the wind and the waves some 200 feet from the pier when he suddenly disappeared.' Though he had been living in Boston, Kelly had been born in Ireland, at Ballinlough in Co Roscommon, where his mother, Catherine, still lived.

It was not until 20th September that Ensign Teulon in LF-2 was able to escort a troop convoy for three hours, from 12 miles North of Malin Head to Rathlin Island. He was in air for four and a quarter hours and reported one destroyer dropping several depth charges,

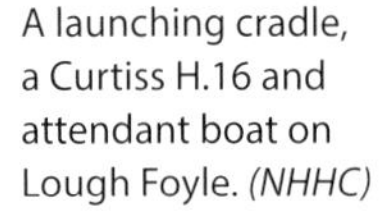

A launching cradle, a Curtiss H.16 and attendant boat on Lough Foyle. *(NHHC)*

supported by three trawlers. He saw nothing and therefore dropped no bombs. Rain prevented sending out a second machine to continue with the escort. On 24th the log noted that an, 'Aeroplane station is being established at Machrihanish for patrol E of Rathlin. Forms part of Luce Bay RAF Group which reports to SNO Larne.' The CO flew again on 26th when LF-2 carried out some practice bomb-dropping and wireless testing. Teulon had reported difficulties with the bomb release mechanism on 20th.

On 4th October LF-3, flown by Ensign George Mongomery, made a heavy landing on return from patrol, following the failure of an engine. He reported, 'the port motor suddenly stopped, while the machine was in a bank, resulting in a side slip from about 1000 to 300 feet; she was then too close to shore to flatten out and make a good landing, so plane landed on water near Warren Point lighthouse with great force, at 90 miles per hour making almost a nose dive, and could just be beached. Hull was damaged beyond repair; engines and planes salved. No casualties.' Some indication of the severity of the weather that autum may be gained from this extract from the log, 'At 9.30 am on 6th October the US Meteorological Station, Malin Head, recorded maximum wind velocity of 99 miles per hour and on 7th October 97 miles per hour. Weather reports copied to 'AMAIR' Queenstown and Malahide Air Station.' Further escorts were flown on 11th and 12th. Another signal extract gives an idea of the scale of the operations at the NAS, 'Director of Stores US Air Station Lough Foyle requires 1200 gallons per week of low grade petrol for motor launch, trucks, powerhouse etc suggest this be supplied through Shell Marketing Co Londonderry and the Co should be informed accordingly so that sufficient reserve be maintained.'

On 13th October LF-2, again flown by Ensign Montgomery, escorted the cruiser HMS *Leviathan*, a sister ship of the unfortunate HMS *Drake*, reporting, 'In air 4 hours 45 minutes, in spite of very bad weather 50 to 60 miles gale and heavy rain.'

A suspected U-boat was attacked by LF-4 on 19th October and two bombs were dropped. George Montgomery's report stated:

The armoured cruiser HMS *Leviathan*. *(Library of Congress)*

'October 19th, 1918
From: George S Montgomery, Ensign, USNRF
To: Commanding Officer
Subject: Report

Left water at 12.37 pm and arrived at Tory Island at 1.45 pm.
Steered NW 30 minutes.
Steered East 20 minutes.
Steered various courses for about 15 minutes, sighting convoy HH72 of 33 vessels at 2.50 pm.
Steered various courses ahead and on flanks of convoy.
At 4.15 pm at a position one mile ahead and two miles on port side of convoy, sighted periscope and well defined wake. Dropped bomb, striking about 30 feet to right of periscope. Signalled to convoy, circled and dropped second bomb, which fell about 10 feet ahead of submarine. Submarine was plainly visible beneath water at this time. Three destroyers stood over and dropped 14 depth charges. Oil and air bubbles appeared.
About 4.35 pm received signal to return to convoy.
Left convoy at 4.45 pm arriving at station at 5.50 pm.
Duration of patrol, 5 hours. 13 minutes. [A record for Irish stations]
All parts of seaplane functioned in a satisfactory manner.

[sgd] Geo Montgomery, First Pilot and JS Thompson, Second Pilot.'

The engineer was CMM Hockridge, the observer CCM Crutchfield and the radioman CM Omholdt.

The crew received a commendation from Vice Admiral Sims. Sadly it would seem that the U-boat was not sunk, as in a letter from Admiralty dated 5th November, stated, 'With reference to the report of an attack on an enemy submarine by a US seaplane from Lough Foyle on 19th October, the Admiralty has classified the result as follows: Chart evidence in this case is indefinite but there is no indication that the submarine was sunk or seriously damaged on this occasion. Classified as possibly slightly damaged.'

Ensign Thompson, from St Paul, Minnesota, was 1st Pilot of LF-4 on 22nd escorting Convoy OLB 14 consisting of five ships. He sent a pigeon message reporting floating mine. On the same day two replacement aircraft, 'left Queenstown at 09.53, will refuel at Wexford.' These H.16s arrived the following day, flown by Ensigns Montgomery and Robb Gover, who hailed from Baltimore. A-3461 was coded LF-6 and A-3465 was coded LF-7. Two patrols were flown on 23rd, fog in the morning prevented escorting convoy HX52, but later on aircraft made sweeps of areas where submarine activity had been indicated by intelligence reports, collated at Queenstown. LF-2, with Parker Teulon as 1st Pilot, had to return with radiator trouble but after this was fixed, carried out a long patrol of four and a half hours duration to eastward. He sighted large and small oil spots and dropped both bombs but with no visible effect. LF-4, flown by James Thompson, patrolled Dungaree Point, Rathlin, Middle Bank, Rathlin, Dungaree Point, where he was forced to land with port engine trouble. A pigeon message was sent and LF-4 was towed in by the motor launch after taxying five miles. Montgomery must have suffered no ill effects from his round trip of 600 miles to Queenstown, as on 24th, he made exceptionally long general patrol in LF-2 covering 'a large

A good view of the flying boat sheds at NAS Lough Foyle. *(NHHC)*

area'. He took off at 10.46 am landing at 16.51 pm, being in the air for six hours five minutes, a new record for Irish stations.

The next day a machine was slightly damaged on take off and had to return to base immediately, when part of its planing hull was damaged. On 26th Montgomery undertook the first operational flight in LF-7, general patrol to westward, remaining in air in spite of rain for four and three-quarter hours. The patrol on 27th, when Thompson in LF-2 escorted convoy HH72 for two hours from five miles NE of Tory Island to 10 miles N of Malin Head, with just under four hours in the air in total, was the last for several days owing to prevailing stormy weather. The log notes that patrol flying would have been attempted had U-boat activity made it necessary, though a test flight was made by LF-6 on 31st with the CO as a passenger.

In an effort to boost morale a Station newspaper was produced – the *Ash-Can Special* for 'the Folks at Home'. The first of only two issues appeared on 2nd November. But a single copy of each has survived – it is a very professionally set out publication, with many interesting adverts from supportive businesses in Londonderry.

When the storm relented on 3rd November, Montgomery and Ensign Charles Ostridge (another son of Massachussets) in LF-6 made the most eventful patrol in the short operational life of Lough Foyle NAS. Their report is as follows:

'Report from: Geo S Montgomery Jr, Ensign USNRF
To: Commanding Officer

Subject: Report of Hostile Submarine Patrol

The Log

8.41	Left water.
9.00	Off Dunagree Point, course 70 decree.
9.30	Off Rathlin Island, sighted two trawlers.
10.00	Off Maidens, course 190°, sighted three trawlers.
10.30	Off South Rock, course 230°
11.00	Off Dundalk Bay, course 240°
11.10	Circles round Dundalk Bay, changing course to 240°
11.45	Off South Rock, course 36°
12.21	Off Carron Point. Course 337° at Lat 55° 10'N Long 5° 38 W Dropped bombs on suspicious oil patch.
1.00	Searched area of Rathlin Island for one hour.
2.00	Off Dunagree Point.
2.45	Landed

Signed: Geo S Montgomery Jr 1st Pilot, Chas Ostridge 2nd Pilot
Endorsed: EH McKitterick'

These bald facts conceal a much more interesting story. It is also of interest to note that the patrol went as far south as Dundalk Bay, evidence of the pilots gaining in confidence.

'From: The Commodore Larne Harbour
To: The Secretary of the Admiralty DASD and Vice Admiral Buncrana
Date: 7th November, 1918 No.3209/73.

Submitted:
1. On Sunday, 3rd November, at 1212, an American Seaplane belonging to Lough Foyle Station dropped three bombs in position approximately 55° N., 5.48 W.

2. It was ascertained that these bombs were dropped on an oil patch observed by seaplane. Airships were detailed to patrol and search area.

3. At 1630, SSZ12 dropped 2 bombs on oil in position approximately 55°.5 N., Long 5.40 W. She spoke to HM Drifter *Flower*, who also dropped depth charges, and who reports considerable quantity of oil rising. Report from *Flower* is attached.

The Pilot of the Airship reports that the oil track he observed appeared to be zigzagging in a southerly direction. Nothing was seen except oil and nothing was heard on hydrophone.
The weather at the time was flat calm; brilliant sunshine.

C Carpendale
Commodore'

The report submitted by *Flower's* Skipper adds further details:

'HMD *Flower*
NAVAL BASE,
LARNE HARBOUR,
3rd November 1918

Chief of Staff
On Sunday, 3rd November I was using Hydrophone on square 15 and I heard two explosions to the Westward, I proceeded at utmost speed in the direction of the explosions. I saw Airship bearing W of me.

After steaming about half an hour, the Airship closed down on me and told me to follow him, which I did. After steaming a short distance the airship dropped a Red Very's light and I found there was oil coming to the surface, so I dropped 2 Depth Charges. After the second Depth Charge exploded a lot of oil came to the surface, also black, oily water, which I think came from a submarine.

I felt sure the submarine was there, so I came round again and dropped another Depth Charge to see if I could bring up anything besides oil.

After I dropped the third Depth Charge, the Airship, whose number was SSZ 12, dropped another Red Light about 400 yards ahead of me, or to the South, and I steamed to where he dropped the Red Light and dropped another Depth Charge. After this exploded, the Airship came down close to me and said I wanted to be about 300 yards further south. I proceeded 200 to 300 yards to the Southward and dropped another Depth Charge. This Depth Charge brought up a large quantity of oil which smelled as if it were a mixture of petrol. I felt sure the submarine was destroyed and I stopped my engines and put out Hydrophone. I used Hydrophone all night until next morning but could not hear anything like a submarine, but oil was rising all night and next morning, so I went and reported it to *The Majesty*.

The position, where I dropped the Depth Charges are, Garron Point bearing WNW and Maidens Light bearing SW and Black Head bearing SSW.

I asked the airship when he came down close if he had seen a submarine and he said yes.

I also fired one Red Rocket signal and 5 Very's Lights.

JT Clarke, Skipper'

On the following day Ensign Montgomery was in the air again, this time flying LF-7 with Ensign Lyman Hodgdon, yet another from Massachussets, as his 2nd Pilot. After being in air about four hours they came down between Ailsa Craig and Mull of Kintyre with a broken cylinder in the starboard engine, and were towed to Campbeltown by the trawler *Goosander*. Montgomery requested by signal that a new engine should be sent over or a vessel to be dispatched to tow LF-7 back to base. On 6th November Ensign Gover carried out a general patrol of three hours duration in LF-6. This would be the final war patrol from Lough Foyle. Meanwhile the rescue tug *St Bees* was dispatched to Campbeltown. On the night of 7th November hurricane force winds lashed the coasts of Ireland and Scotland, with the US Meteorological Station at Malin Head recording a maximum wind speed of 110 mph. LF-7 was washed ashore at Campbeltown and became a total wreck, as signalled by Montgomery, 'Big hole in hull, am dismantling, engines and salvaging parts. Please say where *St Bees* can go alongside to hoist out parts.' He was instructed to go alongside the dock at Londonderry. LF-4 and LF-6 were blown off their moorings on Lough Foyle and onto the beach a short distance to the north, luckily with only slight damage.

On 11th November an official signal was received stating that the Armistice had been signed, 'General quarters sounded and all men assembled in front of hangars and cheered enthusiastically upon hearing the news.' By the time of the Armistice the complement stood at 10 pilots, 10 ground officers and 432 enlisted men.

The last aircraft to fly from Lough Foyle NAS were LF-5 on 12th November and LF-4 on 30th November, both short test flights. A total of 27 patrol flights, 12 training flights

LF5 in front of the hangars at Lough Foyle NAS. *(NHHC)*

and nine test flights, totalling 130 hours and 33 minutes, were made before the war ended. Ten convoys were escorted, at least two U-boats were attacked, with one being claimed as destroyed.

In the second (and final) edition of the *Ash-Can Special*, on 20th November, the CO wrote, 'Going home depends upon the disarmament of the German fleet. It is impossible for me to say officially when that operation will be completed. We must stand by until peace is definitely established, remaining prepared, meanwhile, for instant action, should such again be necessary. Once the word is passed, however, your return to America will be accomplished with the quickest dispatch possible, considering the vast numbers of men in Service who must be transported.'

To celebrate the end of hostilities the Station hosted a 'Peace Banquet and Dance' on 25th November. On 2nd December the first draft of 300 men returned to the USA. The last CO of the station was Assistant Naval Constructor, Lieutenant FJ Wilson, who was appointed on 6th December, having been Executive Officer since 27th August. On 12th December the Station supply of bombs and ammunition was sent to Aghada in Co Cork under guard. The following day, two Packard trucks left for Dublin carrying equipment to be shipped home and work commenced on dismantling the base. One of the US personnel recorded that all the flying boats were set ablaze in January 1919 and all were destroyed within an hour, 'Wings that were sixty feet long disappeared in one and a half minutes.' The Station formally closed on 22nd February 1919.

Had the war continued into 1919 there would have been four USN flying boat stations in Ireland under the overall command of Commander Francis R 'Squinch' McCrary, Naval Aviator No 91 – Aghada (four miles as the crow flies from Queenstown on the south eastern shore of the harbour), Commander JC Townsend, Wexford (at Ferrybank on the River Slaney), Lieutenant Commander VD Herbster, Whiddy Island (in Bantry Bay), Lieutenant Commander Paul J Peyton, and Lough Foyle with a combined total of 84 aeroplanes. Indeed there were contingency plans to base nine flying boat squadrons in Ireland by 1920. The origins of USN aeronautic activity in Ireland go back to a visit made by Captain Hutch Cone, the Head of US Naval Aviation Forces Europe, in October 1917.

Commander Frank McCrary. *(NHHC)*

An inspection tour identified the locations from which the Americans would operate. Little time was wasted. On 22nd October Vice Admiral Sims recommended to the Office of the Chief of Naval Operations in Washington DC that coastal air stations be established at Wexford, Queenstown, Whiddy Island and Lough Foyle.

On 21st January 1918 Cone submitted his formal report, which was, with hindsight, a little optimistic with regard to timescales:

> 'The stations are being built at Bantry Bay, Queenstown, Wexford and Lough Foyle. These stations were decided upon and the locations selected with the object of protecting the shipping approaching and leaving the Irish Sea ports; Bantry Bay, Queenstown and Wexford covering the approaches from the southwest and Lough Foyle covering the approaches from the northwest. These stations will be equipped as follows: Bantry Bay and Queenstown 24 Large Americas each and Lough Foyle and Wexford 18 Large Americas each. Based on what we already know of the capabilities of seaplanes of this type, we can assume that one half of the complement of sea planes will be in operation at all times except in extreme weather conditions and that these seaplanes can scout over areas at least 50 miles to sea from their base. If circles of 50 miles radius are drawn from these seaplane bases, it will be seen that the circles of operation of the three stations on the south coast will overlap and cover the area including the islands and bays from the southwest coast of Ireland well up into the Irish Sea, covering the whole of the straits between St David's Head in Pembrokeshire and Carnshore Point in Co Wexford. So a ship, coming from the southwest and making a landfall at Cape Clear off Co Cork, keeping within 20 miles of the coast, would be under the protection of the seaplanes until they have passed well up into the Irish Sea. The situation at Lough Foyle covers an area of operation from Tory Island off Co Donegal to the Island of Islay off the coast of Scotland. Should it develop that these seaplanes are capable of operating 100 miles from their base, it is seen that a much larger field of operation would be covered, but for the

Hutchinson Cone as a junior officer. *(Author's Collection)*

A flying boat receives attention at Aghada, Queenstown. *(NHHC)*

> present I believe that a 50 mile radius of patrol is as much as can be expected with the number of seaplanes available at the different stations. The delivery of equipment for these stations from America is rather uncertain. It is expected, however, that by May 15th these stations will be operating to one half capacity and by 1st July to full capacity.'

The first stage was the selection of a suitable warehouse in the Port of Dublin to receive the building materials which would be shipped over from the USA and which would become known as the United States Naval Aviation Supply Base, situated at 76 Sir John Rogerson's Quay in Dublin and ready to receive the first shipments in March 1918. These were then

US Naval Airmen at Queenstown. *(NHHC)*

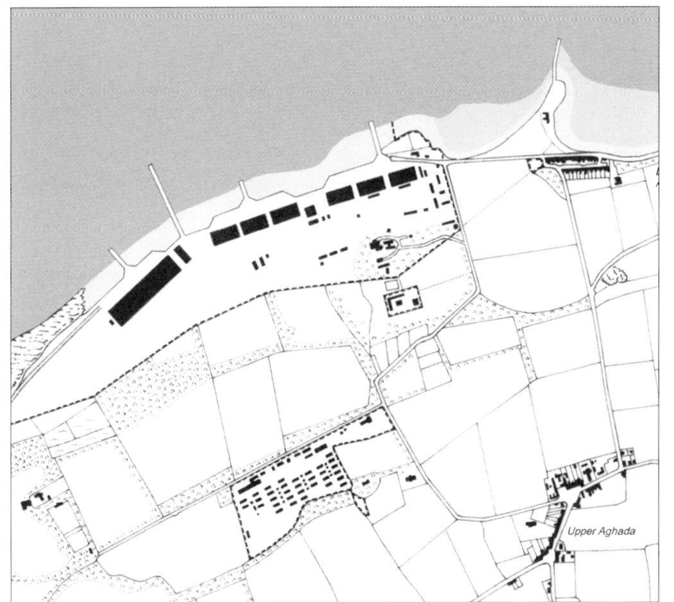

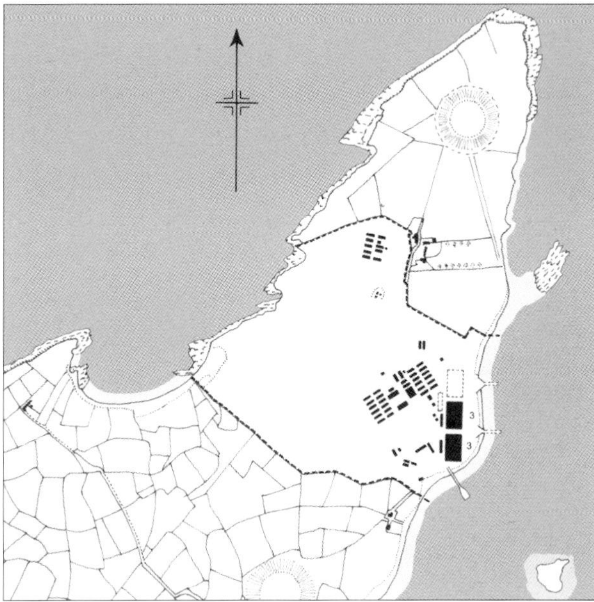

Above left: A site plan of USNAS Queenstown. *(Cross & Cockade)*

Above: A site plan of USNAS Whiddy Island. *(Cross & Cockade)*

forwarded to Queenstown by rail and motor lorry, it having been commissioned on 22nd February 1918 as an assembly and repair base for all naval air stations in Ireland, under the command, in the first instance, of Lieutenant Commander Paul Jones Peyton, with the operational headquarters at Subraon House. The first eight aircraft arrived at Queenstown on 27th June 1918 on board the USS *Cuyama*, with a further ten being brought by the USS *Kanawha* a month later, out of an eventual total of 38.

The first test flight from Aghada was made on 3rd August by Ensign J William Lancto, who was from up-state New York. As it was, Queenstown (Aghada), Wexford and Whiddy Island were operational in October and November with up to half a dozen H.16s at each location, with flying having commenced at the latter two sites in September. Station personnel at

USNAS Whiddy Island. *(Author's Collection)*

Queenstown devoted much of their time to assembling machines, testing them and ferrying them to other bases. Mechanical difficulties with the engines were common. However, on 16th October 1918, a flying boat from Wexford, crewed by Lieutenant John F McNamara, Lieutenant George Shaw and NAS Queenstown Ensign James Roy Biggs dropped bombs on a submarine 'which then surfaces at irregular intervals and eventually disappears' leaving large quantities of debris and oil, but the Admiralty assessed it as only 'probably' seriously damaged. McNamara and Shaw were both awarded the Navy Cross for their service overseas.

Walford Anderson *(Author's Collection)*

Several aircraft were lost due to mechanical failure or pilot error, A-1072, A-1074, A-1075, A-3550, and A-4054. One crew member was killed, a radio operator, Petty Officer Walford August Anderson, on 22nd October 1918. He was returning from a patrol in A-1072 when it crashed into the sea off the coast of Whiddy Island, which had been commissioned on 4th July 1918. His friend and fellow wireless telegraphist, Harvey Wilson, later wrote:

> 'My good friend, Walford Anderson, came running down to the shore. "Harvey", he called, "there is a ship being attacked by a German sub and they are getting your plane ready to go. To avoid delay, if you don't mind, I'll go in your place." I signalled it was OK with me and he ran and boarded the flying boat. It took off down Bantry Bay and disappeared from our sight.
>
> Back ashore, and supper over, I went to the Wireless Shack to listen for how my plane with Walford in it could be progressing. It was out longer than expected, and came in to Bantry Bay as it was getting almost dark. There was no wind and the bay was smooth as glass. The pilot and second pilot both, sort of puzzled by the smoothness of the bay, said it was like a mirror. They misjudged their height, or altitude since the mirror effect gave them the impression they were much higher and

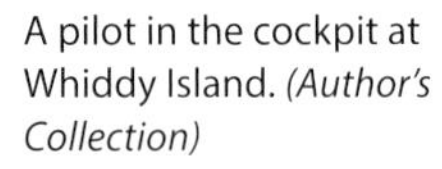

A pilot in the cockpit at Whiddy Island. *(Author's Collection)*

they brought the plane down too steeply. Noticing the mistake, they tried to level its flight, but not quickly enough. It sort of dived into the water, but so violently it tore the starboard Liberty motor off its mounting. Walford's wireless set being a bit forward but directly under that heavy 12 cylinder motor, it crashed through the body cover and hit Walford on the head. He was killed, they said, instantly. My best friend had died in my stead. I should have been in that plane where he sat in my seat.'

The Aero Nut.

— On dit —

Vol. I. Queenstown, December 31st, 1918. No. 12

PRICE: Kind Words. We will do all the knocking necessary.

ADVERTISING RATES SECRET.

A Happy New Year.

Our Exit Made Easy With Encomiums By Stations Officials.

Commander McCrary, Commander Townsend, and Lieutenant Jalbert Write Articles for Last Issue of "Nut."

"Z.Y.S."—"WELL DONE."

S. Naval Aviation Base, Queenstown, Ireland, Dec. 26th, 1918.

the Editor of "The Aero-Nut."

In the last issue of the "Aero-Nut" will save me space in which to bid good-bye the officers and men of the U.S. Naval Stations, Ireland. It is the pride and of the Navy that its officers and men perform any work asked of them, and its

(Continued on Page 3)

"The Aero Nut" Makes Its Curtain Bow With This Twelfth Number.

Saturday and Robinson Backbone of Paper. Lieutenant Stitt Originator of Idea.

MEMENTO.

With this issue "The Aero-Nut" passes out of existence. Its life has been a short and merry one.

If we have in any way pleased you, we are well repaid for our efforts; if not—well we will go without pay. It is the customary thing when making an exit such as we are doing to humbly beg the pardon of any who may have taken offence, but we are not custom bound, and, therefore, offer no apology

We wish to thank all those who have aided us directly and indirectly; those who have

The Aero Nut. (via Ronnie Vogt)

Queenstown NAS produced a weekly illustrated journal, *The Aero Nut*, which ran for 12 issues between the start of October and the end of December. Its farewell message was as follows, 'Adieu Erin, during the months we have spent with you, with your generous goodwill, sincere sociability, and warm hospitality, you have won a place in our hearts. We shall always cherish the most pleasant memories of our sojourn here, and our earnest hope is that such shall be your memories of us.'

The Executive Officer of Wexford NAS, Lieutenant Clarence B Tillotson, edited and compiled a souvenir booklet, assisted by Chief Carpenter's Mate Charles E Riley and Carpenter's Mate 1st Class Gordon E James. The crew roll therein contains the surprising total of 444 names, with a dozen pilots supported by intelligence, pay and supply, medical, meteorological, engineer and radio officers, quartermasters, gunners, boatswains, coxwains, seamen, machinists, electricians, carpenters, yeomen, storekeepers, pharmacists, stewards, ship fitters, coppersmiths, blacksmiths, cooks, bakers, painters and firemen. The majority were enrolled for construction work. It was noted that the CO arrived on 28th March 1918 on a site which rapidly became a sea of mud as men worked from 5.00 am to 9.00 pm daily in incessant rain to create the base. Herbster was Naval Aviator No 4 and was a strong advocate of hard work, strict discipline, athletics and meticulous preparation. At one stage the local contract

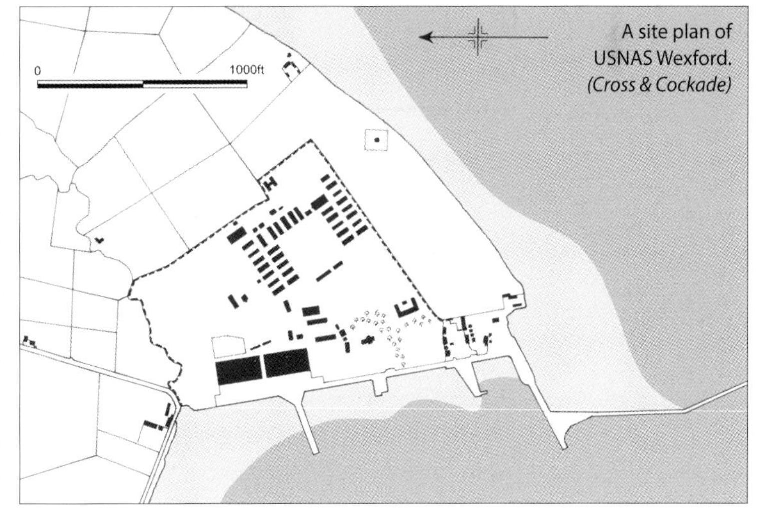

A site plan of USNAS Wexford. *(Cross & Cockade)*

Wexford NAS. Lieutenant Commander Victor Herbster is on the far left. *(Author's Collection)*

workforce staged a strike which was quickly halted when it was seen that the USN could carry on with the construction work themselves.

The station was situated on the northern shore of Wexford Harbour. From a strategic point of view it lay at the southern entrance of the Irish Sea between Carnshore Point and the Tuskar Lighthouse. It was commissioned on 2nd May 1918. The civilian services were dispensed with in August. The first four aircraft arrived from Queenstown on 18th September, with flying commencing the following day. Only five of the total planned complement of 18 flying boats actually arrived before hostilities ceased. Four attacks were made on suspected

An aerial view of USNAS Wexford. *(Author's Collection)*

U-boats, with No 1079 dropping two bombs on an 'enemy submarine' on 21st September, as reported in a signal message from 'Aeronautics Wexford'. In all 98 patrols were flown over a period of eight weeks. The authors concluded their brief account of activities at Wexford with the claim that with its, 'warm, comfortable barracks and attractive mess hall, where good food was plentiful' it was 'the neatest, most up-to-date and best station in Europe' adding, 'cleanliness, neatness and pride in the station were the watchword.' Commander McCrary made a weekly report to Admiral Bayly, giving statistics of all missions flown. He drew the Admiral's attention to two patrols made just before the Armistice, evidence of the US airmen's growing reach:

Flying boat W3 at Wexford. *(via Frank Brophy)*

> 'Special attention is invited to two endurance flights made by Seaplanes from Aghada on November 9, 1918. Seaplane No 4045, piloted by Ensign Phillip H Gadsden, first Pilot, and Ensign David J Wood, Second Pilot, carrying full equipment, including two bombs, made a flight lasting 8 hours and 3 minutes, and on the same date Seaplane No 1048, piloted by Ensign Martin J Dwyer, First Pilot, and Ensign George L Compo, Second Pilot, made a flight carrying full equipment, including two bombs, lasting 9 hours and 37 minutes.'

A grand total of 761 flying hours was achieved by the four US Naval Air Stations, covering 48,000 patrol miles, 19,000 miles at Wexford, 15,000 at Queenstown, 10,000 at Lough Foyle and 4000 at Whiddy Island. Seven suspected submarines were bombed. Flying conditions were often atrocious. By April 1919 the USN had vacated all of its Irish bases.

A Curtiss H-16 on the river at Wexford. *(Frank Brophy)*

Chapter 11

The US Navy around the rest of Ireland, 1917–18

Top: Admiral Sir Lewis Bayly RN. *(Author's Collection)*

Above: US Secretary of the Navy Josephus Daniels. *(NHHC)*

'In April I was sent for by the Admiralty,' Admiral Bayly later recalled, 'and they told me that in May some United States destroyers were coming over to help us. They would be put under my orders and the Admiralty hoped I would be 'nice' to them. I was introduced at the Admiralty to Sims, the American Admiral. On my way back to Queenstown I wondered how to be 'nice' to them, and finally decided to treat them exactly the same as I treated the British.'

Rear Admiral William Sowden Sims had been sent over to England in March as the Senior Representative of the US Navy. With the declaration of war he assumed command of the US naval forces which would soon be on their way to the war zone and was promoted to Vice Admiral. He was a close and trusted friend, later writing, 'There were no secrets of the British navy which were not disclosed to their new American ally. This policy was in accordance with the broad-minded attitude of the British Government; there was a general desire that the United States should understand the situation completely, and from the beginning matters were discussed with the utmost frankness.' He quickly grasped the essential points that as many US destroyers should be sent to European waters as soon as possible – to Queenstown in particular, that they should be placed under the existing command structure and that the convoy system was the key to stemming the loss of merchant shipping, writing, 'The situation which confronted us in April, 1917, was one which demanded an immediate and powerful offensive; the best way to protect America was to destroy Germany's naval power in European waters and thus make certain that she could not attack us at home.'

On 14th April Secretary of the Navy Josephus Daniels had issued movement orders for Division Eight, Destroyer Force, US Atlantic Fleet, to deploy to European waters, 'to assist naval operations of the Entente Powers in every way possible' and proceed to Queenstown.

The first six destroyers of the US Navy arrived in Ireland on 4th May 1917, with eventually up to 37 destroyers being based at Queenstown on convoy escort duties, 'being selected as a base of operations on account of its proximity to the focus of traffic lanes to the waters of Great Britain and northern France.' They were met off the Daunt's Rock Lightship by Lieutenant E Keble Chatterton

RNVR, commanding ML 181, 'on a balmy, sunny May morning with a calm sea and a certain amount of Atlantic swell.' Vice Admiral Sims gave his impressions of the scene:

ML181 going out to meet the US destroyers on 4 May 1917. *(Author's Collection)*

> 'At almost the appointed hour a little smudge of smoke appeared in the distance, visible to the crowds assembled on the hills; then presently another black spot appeared, and then another; and finally these flecks upon the horizon assumed the form of six rapidly approaching warships. The Stars and Stripes were broken out on public buildings, on private houses, and on nearly all the water craft in the harbour; the populace, armed with American flags, began to gather on the shore; and the local dignitaries donned their official robes to welcome the new friends from overseas. One of the greatest days in Anglo-American history had dawned, for the first contingent of the American navy was about to arrive in British waters and join hands with the Allies in the battle against the forces of darkness and savagery. The morning was an unusually brilliant one. The storms which had tossed our little vessels on the seas for ten days, and which had followed them nearly to the Irish coast, had suddenly given way to smooth water and a burst of sunshine. The long and graceful American ships steamed into the channel amid the cheers of the people and the tooting of all harbour craft; the sparkling waves, the greenery of the bordering hills, the fruit trees already in bloom, to say nothing of the smiling and cheery faces of the welcoming Irish people, seemed to promise a fair beginning for our great adventure.'

Destroyers under the command of Commander Joseph K Taussig USN, arriving at Queenstown, on 4 May 1917. *(NHHC)*

The USN historian, Captain Dudley W Knox adds more prosaically:

> 'The duty of escorting convoys was extremely arduous. The small vessels had to keep the sea for long periods and maintain the same speed as the convoy regardless of weather conditions. Many convoys had to be met as far as 300 miles from the coast. The great extent of the ocean combined with the comparatively few [about 12] submarines which the Germans could maintain continuously on station prevented frequent attacks by enemy submarines. Many escort vessels went through the entire war without seeing a hostile submarine, but this was due in part to the fact that the submarines preferred to leave the protected convoys alone and to expend their efforts in the less dangerous work of attacking single ships of which one or more usually straggled from each convoy.'

Commander Joseph K Taussig USN. *(NHHC)*

Admiral Bayly's directive to Commander Joseph K Taussig the commander of Division Eight, had the merit of being short, simple and direct: (1) to destroy U-boats (2) to protect and escort merchantmen (3) to save the crews and passengers of torpedoed ships. When the Admiral asked Taussig on the evening of his arrival after a tough nine days crossing the Atlantic, 'When will you be ready to go to sea?', Taussig replied, 'We are ready now, sir, that is, as soon as we finish refueling. Of course you know how destroyers are, always wanting something done to them. But this is war, and we are ready to make the best of things and go to sea immediately.' Sims vividly described that first encounter:

> 'It is doing no injustice to Sir Lewis to say that our men regarded this first meeting with some misgiving. The Admiral's reputation in the British navy was well known to them. They knew that he was one of the ablest officers in the service; but they had also heard that he was an extremely exacting man, somewhat taciturn in his manner, and not inclined to be over familiar with his subordinates, a man who did not easily give his friendship or his respect, and altogether, in the anxious minds of these ambitious young Americans, he was a somewhat forbidding figure. And the appearance of the Admiral, standing in his doorway awaiting their arrival, rather accentuated these preconceptions. He was a medium-sized man, with somewhat swarthy, weather-beaten face and black hair just turning grey; he stood there gazing rather quizzically at the Americans as they came trudging up the hill, his hands behind his back, his bright eyes keenly taking in every detail of the men, his face not showing the slightest trace of a smile. This struck our young men at first as a somewhat grim reception; the attitude of the Admiral suggested that he was slightly in doubt as to the value of his new recruits, that he was entirely willing to be convinced, but that only deeds and not fine speeches of greeting would convince

> him. Yet Admiral Bayly welcomed our men with the utmost courtesy and dignity, and his face, as he began shaking hands, broke into a quiet, non-committal smile; there was nothing about his manner that was effusive, there were no unnecessary words, yet there was a real cordiality that put our men at ease and made them feel at home in this strange environment.'

The next morning the Admiral addressed them again, 'Gentlemen, the Admiralty are afraid I shall be rude to you. I shan't if you do your work; I shall if you don't.' The US officers greatly appreciated his frank speaking and the foundation stone of a memorable relationship was thus laid. Indeed the austere and outwardly stern C-in-C came to be known behind his back by his officers and men as 'Uncle Lewis.'

On 19th May, Bayly again spoke to the commanding officers of the US destroyers. He told them that in two days time the destroyers would head out on war patrols and passed on guidance to the American officers. His remarks were very direct and to the point:

> 'When you pass beyond the defences of the harbour you face death, and live in danger of death until you return behind such defences. You must presume from the moment you pass out that you are seen by a submarine and that at no time until you return can you be sure that you are not being watched. You may proceed safely, and may grow careless in your watching; but, let me impress upon you the fact that if you do relax for a moment, if you cease to be vigilant, then you will find yourself destroyed, your vessel sunk, your men drowned.'

USS *Ericsson* (NHHC)

They were soon in action. On 21st May USS *Ericsson*, (of Division Seven, which had arrived on 17th May) while on patrol in the Western Approaches, sighted a German submarine running on the surface engaging two sailing vessels, a Russian and a Norwegian, with its deck gun. *Ericsson* closed on the submarine and opened fire, also launching a torpedo – the first to be fired at the enemy by a US warship. Seeing the destroyer's approach, the submarine dived below the waves and sank both sailing vessels with its own torpedoes. *Ericsson* searched for the U-boat but could not locate it and then proceeded to pick up the surviving crew of the sunken vessels, disembarking the survivors at Queenstown, after which *Ericsson* returned to her patrol. Bayly noted:

> 'After the US naval officers had learned their way about I treated them exactly the same as the British. The two Navies were mixed up in escorts, etc, when necessary,

Depot ship USS *Melville* tending destroyers and submarine chasers at Queenstown in 1918. Flying at the peak of her mainmast is the three-star flag of Vice Admiral William S Sims. *(NHHC)*

> in every case the senior officer taking charge. It worked perfectly, though the British ships were mostly 16-knot ships and the US were 30-knot destroyers. In fact, on one occasion when the escort had a long way to go to meet a convoy, and the senior ship was a British sloop with a very able Commander, he made a signal to the senior US destroyer that on account of their superior speed he would like him to take command. A very good proof of the excellent spirit existing between them.'

The needs of the US ships were met by the dockyard and also by two well-equipped depot ships USS *Melville*, Captain JR Poinsett Pringle and USS *Dixie*, Captain HB Price. These were anchored in Monkstown Bay, which was in Passage West on the west bank of Cork Harbour about four miles upriver from Queenstown and six miles downriver from Cork.

The Naval Centre at Queenstown was undoubtedly the focal point for the collection and dissemination intelligence for the entire operation around the coast of Ireland, receiving reports from Larne, Kingstown, Galway, Buncrana, Berehaven, Killybegs and Rosslare, as well as all the Coastguard units, lighthouse keepers and wireless telegraphy intercept stations; and further afield from Brest, Milford Haven, Devonport, the Scilly Isles and Lands End. By this means a picture was built up and maintained of U-boat activity, sightings, sinkings, escorts, convoys, incidents, attacks by RN and USN ships and aircraft – a daily record of dates, times and positions, all plotted on Admiral Bayly's chart table, which had previously been the Admiralty House billiard table.

By the week ending 3rd July 1917 four convoys comprising the First American Expeditionary Force had been safely escorted on the final stages of their Atlantic crossing

Above: Barely afloat, after a collision with HMS *Zinnia*, the USS *Benham* is brought alongside the USS *Melville* and the USS *McDougall. (Author's Collection)*

Above right: Depth charges being loaded onto an American destroyer at Queenstown. *(Author's Collection)*

by US destroyers attached to the Irish Command.

There were some teething problems, however, such as occurred on 20th August when the sloop, HMS *Zinnia*, collided with the destroyer, USS *Benham*, nearly sinking the latter. *Zinnia* later towed *Benham* to Queenstown for repairs.

The process of learning was dynamic and empirical. Admiral Bayly convened a conference in Queenstown, on 8th September, of British and American commanding officers of all sloops, destroyers and repair ships/tenders under his command to compile suggestions for the best means of sinking submarines and protecting shipping. Vice Admiral Sims passed the conference conclusions, 'an array of technical and tactical operational and formational suggestions', to the Admiralty for consideration.

What was it like to serve in one of these 'long-legged, oil-burning' destroyers?

An article by Vice Admiral Walter S Delany, who served for fourteen months in the USS *Wilkes* at Queenstown supplies some of the answer:

> 'They were 1100 ton, four-stackers with four 4-inch single-purpose guns and four triple-tube 21-inch torpedo mounts. They were not equipped with any special anti-submarine weapons except, initially, the hand-thrown 50-pound depth charge, or "ashcan" as it was commonly called. There were no arrangements to launch them, and ships picked the strongest man in the crew to heave them over the stern when attacking a suspected periscope, oil slick, or unsuspecting whale. The charges were made in two parts which separated when they hit the water. One part acted as a float and the other as the explosive charge. The depth to which the charge sank was controlled by a line hand-wound on a spindle. When the charge sank to that depth, it was triggered and blew up. Later on, when 300-pound charges fitted with variable depth settings were developed, they were stowed in two fore-and-aft tracks secured to the after open deck. They could be rolled overboard in an attack by releasing them singly from a hand operated pull on the bridge. Y-guns were an added and later installation for projecting these charges to safe distances

abeam. In addition, "fighting lights" were installed. These were red and green lights which could be flashed together as an emergency recognition signal. They were the initial IFF [identification, friend or foe] installation. These installations, as well as changes to the forward bulkhead of bridges to provide enclosed protection for the watch standers against the foul weather, were usually accomplished at a dockyard in Liverpool. USS *Wilkes* had its hull painted with a new camouflage design by Henry Bittinger, the Navy specialist in the Bureau of Ships. His idea was to paint out shadows such as overhangs, the under side of guns, etc, with merging tones of gray and white paint. This was not acceptable in the war zone where the "dazzle" design replaced it.

When on patrol or escort duty periscopes were to be shelled and not rammed to avoid decoys attached to mines. Ships were always to zig-zag and never to steam at less than 13 knots. A British signalman always spent one month on board each arriving US destroyer to train its seamen in the British system. Ships had orders to keep their radio room informed of their position every two hours. In case of an accident, the operator would send the ship's position without further orders.

Everything did not always come up roses for all hands when they came to Queenstown. Those familiar with destroyers in the old Navy know that a real characteristic of a good leading Chief was his ability to forage. His store rooms always had many 'you never know when you might need this' items which did not appear on the books. This was especially true when the word got around that their ship was getting ready to go to the war zone overseas. The store rooms were filled with spare parts, extra mooring lines, tools, and other items that 'just happened' to find their way on board. These carefully laid plans, however, went overboard when the destroyers arrived at Queenstown. To the complete disgust of these far-sighted Chiefs, they found a station order which required ships to land for common

USS *Wilkes (NHHC)*

use all except a minimum quantity of stores, spare parts, etc, for the specific purpose of lessening the ship's draft. To decrease the over-horizon range at which the ships could be seen, the topmast had to be housed and the radio antennae replaced.

A US destroyer alongside a depot ship at Queenstown. *(Author's Collection)*

Admiral Bayly wrote, "It is a great tribute to both British and US captains that I never had a signal 'Request instructions'. Their orders were to do what they considered the best thing to do when in doubt, and very well they did it." He added, "I always saw the captains of escort ships of both Navies the morning after their return to harbour and the morning before they sailed: the former because there was always something to learn from them, the latter to make sure that they understood what they had to do."

They were lively ships in winter storms. A tale around Queenstown relates to one destroyer that picked up the captain and two crewmen in a boat from a torpedoed tramp. The destroyer skipper was concerned that the merchant skipper would not stay in the wardroom but continued to sit on a camp stool lashed to the engine room hatch amidships. In conversation in this connection, the merchant skipper is reported to have said he had been going to sea for 30 years but had never been on such a ******* active ship before. He had been all over the destroyer and found the camp stool to be the most comfortable and that's where he was staying. The wardroom messman brought him his ration of sandwich and soup which was the menu for such weather. A Liverpool pilot endorsed this somewhat uncharitable opinion, "The navigator of the destroyer with the pilot on board mentions finding the pilot standing in the passageway from the wardroom to the outside deck, where he was going to the bridge for his 4–8 am watch. The pilot, with his spread feet and arms against the bulkheads, was a depressed looking person. The navigator, trying to make conversation said, 'How goes it, pilot,' and he could not be mistaken about the pilot's feeling when after some expressive words he concluded with, 'Every bone in my bloody body aches from 'olding on in this ******* ship.'"

Of interest to today's Navy is the fact that there was practically no paperwork other than operational orders and official reports. Ships operating on a five-day out and two or three-day in-port schedule had no time for paperwork. Unlike the customary work requests to be "submitted in quintriplicate for approval" and then await work commencement, when returning ships neared the entrance to Queenstown Harbour they would exchange recognition and distinguishing signals by flashing light with the signal station on the beach. They also requested any important repairs and supplies needed. Before entrance, the destroyer would have been assigned berths at the oil jetty or numbered mooring buoy without any inquiries from the ships. As soon as the ship was secured, all unnecessary fires were allowed to die out, and machinery needing repairs was disabled. It was known that the ship, except for an emergency, would be in port for three days. The convoy schedules and work was planned accordingly. Incoming destroyers were fueled in pairs at the jetty. As the fueling was completed, the ships were moved by tugs to their next mooring. Even before the destroyers had secured in their assigned berths, motor launches from the tenders mothering the destroyers would be alongside, with repair officers, fresh provisions, mail, and necessary repair and replacement parts. The crews of the tenders were on duty around the clock to provide such services. Since each destroyer averaged about 6000 miles per month under all kinds of weather, these hard working and devoted tender crews certainly rate the high regard in which they were held by all the operating destroyer personnel. The dockyard officers were told that there would be no defect lists and that they must keep an account of the money expended on British and US ships. This was done and in 1919, when the war was over, the Admiralty sent a statement over to the USA of accounts which had been paid.'

An impression of the effect the American and RN escorted convoys had on the U-boats and their crews may be gained from this account, first published in *Raiders of the Deep* in 1929:

'In August Kapitänleutnant Otto Hersing demonstrated that an experienced U-boat skipper could still obtain results provided he pressed his attack with resolution and determination, even commanding a veteran U-boat of 1913-vintage. U-21 encountered a convoy fifty miles off the south-west tip of Ireland. It was, 'one of those rare sparkling days with hardly a ripple on the sea' and the surface of the water was so glassy that Hersing did not dare to use his periscope for more than a few moments at a time. There were fifteen ships in the formation with six destroyers on either flank and two more, one half-a-mile ahead, the other a similar distance astern. It looked an impregnable defence but Hersing was too old a hand to be deterred simply by a show of strength. Diving between two of the flank destroyers and poking his periscope up just long enough to estimate the course and speed of his targets Hersing fired two shots. Then, diving to 40 metres he waited results. Both torpedoes scored direct hits, though it

A depth charge from a US destroyer explodes. *(Author's Collection)*

would appear from the records that no vessel was actually sunk. Retribution followed swiftly. Guided by the bubbling tracks of the torpedoes the destroyers homed on their hidden enemy. A veritable deluge of depth-charges rained down on the lurking submarine. Hersing and his crew were shocked and shaken by the violent explosions that followed. "Every square yard of water was being literally peppered with depth bombs. They were exploding on every side of us, over our heads, even below. The destroyers were timing them for three different depths – ten metres, twenty-five metres, and fifty metres … They were letting us have them at the rate of one every ten seconds." A near miss shattered the lights and Hersing was convinced that they were finished. But the reports coming into the control room confirmed that there was no damage and, gritting his teeth, he twisted U-21 first one way and then the other to escape his tormentors. "The sound of propellers followed us wherever we went and the bombs continued their infernal explosions. The U-21 shivered with each detonation – and so did we." The hunt continued for five nerve-wracking hours and then, satisfied that they had destroyed the submarine, the destroyers moved off to rejoin the convoy. After such an experience it was small wonder that few U-boat captains tackled an escorted convoy twice.'

Returning to Vice Admiral Delany's account, he greatly appreciated Admiral Bayly's attitude to legal matters and Courts of Enquiry and cited as an example:

USS *Shaw* (DD-68) was escorting HMS *Aquitania* with 8000 men aboard when the Shaw's rudder jammed while zigzaging. Unable to avoid collision, *Aquitania* rammed *Shaw*, slicing a 90 foot section from her bow, toppling her mainmasts, and destroying her forward boilers. USS *Shaw* backed 40 miles into port to be repaired and sail again. (NHHC)

'As an indication of the decisiveness of the Admiral in matters of this nature, it is interesting to cite the finding in the case of the collision between the USS *Shaw* and the British liner *Aquitania* while the latter was under escort in the English Channel. The helm of *Shaw* jammed at almost full left rudder and turned her across the big ship's bow. Result, the *Shaw's* bow was cut off. The Admiral's recommendation was typically to the point:

To: The Secretary of the Admiralty
Date 13th October 1918

HMS AQUITANIA & USS SHAW – Collision
Herewith is forwarded a report by the Commanding Officer of the USS SHAW of collision with HMS AQUITANIA on 9th October 1918. The USS SHAW is at Portsmouth.

The report shows the coolness, knowledge of

the ship, and quickness of decision which I should have expected of Commander Glassford and a ship commanded by him.

As far as a mixed Court of Inquiry usually formed here when British and U.S. ships are concerned, it does not seem to me to be necessary in this case, as I do not see how HMS AQUITANIA would have known that USS SHAW's helm had jammed, or how she could have avoided her.

Lewis Bayly
Admiral
Commander-in-Chief'

As can be seen, Bayly's letter was brief to the point of terseness. This tended to be the laconic 'house style' of the Royal Navy. Contemporary US Navy reports are much more detailed and indeed verbose.

One of the Commanding Officers at Queenstown would gain fame during the naval war in the Pacific 25 years later, Commander William Frederick 'Bill' or 'Bull' Halsey, serving in the USS *Benham*, *Duncan* and *Shaw*.

At rest in Queenstown

On shore the leisure requirements of an eventual total of 7000 men were met by the establishment of the US Sailors' Club, which opened on 25th August and contained a stage, a film projector, kitchen, restaurant, canteen, pool-room, library, dormitory and showers.

British ratings were made honorary members. A recreation hut was erected on Haulbowline Island for British sailors by the YMCA. Officers of both nations could repair to the Royal Cork Yacht Club, while the walled-in 'Sloop Garden' at Admiralty House, offered respite for commanding officers. The COs of the sloops, which were named after flowers, took a pride in adding a suitable plant to the herbaceous border.

Many also appreciated the hospitality, kindness and charitable good works of Miss Violet Voysey, the Admiral's niece, who acted as hostess. Commander Taussig later wrote about

Below: The US Navy Men's Club at Queenstown. *(Author's Collection)*

Below right: A game of pool in the US Naval Men's Club, Queenstown. *(Author's Collection)*

a sporting activity that introduced the Americans to a very British pastime:

A Flower Class Sloop in dazzle-paint docked at Whiddy Island. *(Author's Collection)*

> 'At four o'clock Captain Pringle picked me up and we went to the Yacht Club where we were joined by Fairfield [USS *McDougal*] and Zogbaum [USS *Davis*]. From there to the Admiralty House where we had tea with Admiral Bayly, Miss Voysey, and a Miss French. After tea we played 'cricket' or really 'tip and run,' I think, is the proper name. It was played on the front lawn, a tennis ball and small stick being used. Each one batted and bowled in turn, Admiral and ladies included. It was not wildly exciting, but there was plenty of good exercise in it for one who has not been taking any regular exercise. I expect to be stiff from it tomorrow.'

They must have enjoyed it as it became a regular occurrence. Sims adds that Bayly:

> '... was a man of wiry physique and a tireless walker. Indeed the most active young men in our navy had great difficulty in keeping pace with him. One of his favourite diversions on a Saturday afternoon was to take a group on a long tramp in the beautiful country surrounding Queenstown; by the time the party reached home, the Admiral, though sixty years old, was usually the freshest of the lot. I still vividly remember a long walk which I took with him in a pelting rain; I recall how keenly he enjoyed it and how young and nimble he seemed to be when we reached home, drenched to the skin. A steep hill led from the shore up to Admiralty House; Sir Lewis used to say that this was a valuable military asset, it did not matter how angry a man might be with him when he started for headquarters, by the time he arrived, this wearisome climb always had the effect of quieting his antagonism. The Admiral was fond of walking up this hill with our young officers; he himself usually reached the top as fresh as a daisy, while his juniors were frequently puffing for breath.'

A little spaniel, Patrick, was indispensable part of the Admiral's 'family'. The three were entirely devoted to one another. Sims wrote, 'I have never seen any object quite so crestfallen and woe-begone as this little dog when either Miss Voysey or the Admiral spent a day or two away from the house.'

The US officers were encouraged to engage in social interaction by Bayly, as Sims recalled:

> 'He even took a certain pride in his ability to comprehend the American joke. One of the regular features of life at Queenstown was a group of retired British officers – fine, white-haired old gentlemen who could take no active part in the war but who used to find much consolation in coming around to smoke their pipes and to talk things over at Admiralty House. Admiral Bayly invariably found delight in

Admiral Sir Lewis Bayly on board the USS *Cushing* at Queenstown. *(Author's Collection)*

encouraging our officers to entertain these rare old souls with American stories; their utter bewilderment furnished him endless entertainment. The climax of his pleasure came when, after such an experience, the old men would get the Admiral in a corner, and whisper to him: 'What in the world do they mean?"

Comfort was also given to the victims of ships sunk by the enemy brought into Queenstown. Sims again:

'In a large hall in the Custom House at the landing the Admiral kept a stock of cigarettes and tobacco, and the necessary gear and supplies for making and serving hot coffee at short notice, and nothing ever prevented him and his people from stationing themselves there to greet and serve survivors as soon as they arrived often wet and cold, and sometimes wounded. Even though the Admiral might be at dinner he and Miss Voysey would leave their meal half eaten and hurry to the landing to welcome the survivors. The Admiral and his officers always insisted on serving them, and they would even wash the dishes and put them away for the next time. The Admiral, of course, might have ordered others to do this work, but he preferred to give this personal expression of a real seaman's sympathy for other seamen in distress. It is unnecessary to say that any American officers who could get there in time always lent a hand. I am sure that long after most of the minor incidents of this war have faded from my memory, I shall still keep a vivid recollection of this kindly gentleman, Admiral Sir Lewis Bayly, KCB, KCMG, CVO, RN, serving coffee to wretched British, American, French, Italian, Japanese, or negro sailors, with a cheering word for each, and afterward, with sleeves tucked up, calmly washing dishes in a big pan of hot water.'

Delany adds:

'With the small size of the town, lack of recreation facilities and even religious differences, liberty was a problem in Queenstown, and personnel were permitted

Below: Rescue of crew of the a merchant ship after it was torpedoed. Picture taken from deck of USS *Beale* (DD-40). *(NHHC)*

Below right: Part of the crew on board USS *Beale* after their rescue. *(NHHC)*

to go to Cork by train. It soon became apparent that this was creating difficulties for all hands. The [nearly 8000] US sailors had no other place to spend their money which was of sizeable amounts compared to that earned by local males. Obviously, the American blue jackets became popular with girls in Cork. The reverse was true of their boy friends. The result was free-for-alls in Cork and an order prohibiting British and US officers and men to go within three miles of Cork on any pretence. It was learned that Cork suffered a loss of about £4000 per week as a result of that order.'

The Lord Mayor of Cork was the leader of a deputation to the Admiral, asking him to rescind the order. They received a fair but brief hearing which was answered in the negative when they could not guarantee the good behaviour of their citizens. As a final gambit the Mayor asked if the sailors could be allowed into town until 4.00 pm. The Admiral replied, 'in acid tones' that he could not agree a proposal which would permit the US personnel to spend money in the shops but then, 'leave Cork just as they are beginning to enjoy themselves.' At the foot of the hill the Mayor was asked if he had been successful. He responded, 'I did not indeed. But by the grace of God I left through the door and not by the window.'

Operational matters

By the late summer of 1917 there were 35 US destroyers, 15 RN sloops, nine Q-ships and HMS *Adventure* based at Queenstown, a total of 60 warships, to which could be added a host of Motor Launches (ML), trawlers and drifters. The MLs' usual routine was eight days on patrol and two days in Queenstown, with a four day spell in Queenstown every three months. Different harbours and bays would be visited on each patrol. Without radio the MLs reported at the various coastguard stations, which passed the information back to Queenstown. In fine weather the MLs would usually anchor in the vicinity of a coastguard station overnight, conducting a hydrophone watch. In bad weather they would anchor in one of the small ports or bays for the night. In their two days at Queenstown the ships would replenish with fuel, food and stores, complete any maintenance required and have a few hours leave.

The destroyer USS *Cassin*, Lieutenant Commander Walter N Vernon, was operating off Co Waterford out of Queenstown.

On 15th October 1917, *Cassin*'s lookouts sighted U-61 which immediately submerged and began to flee. The pursuit ensued for an hour until Kapitänleutnant Victor Dieckmann decided to engage the US warship. He turned about and surfaced to line up for a shot and fired his last torpedo. Gunner's Mate First Class Osmond Ingram noticed the incoming wake, he quickly ran over to the depth-charge gunners and ordered them to shoot charges in U-61's direction. The torpedo struck the destroyer aft on the port side, nearly blowing off her rudder, before the depth-charge attack could be launched. Ingram was killed in the explosion. The American destroyer began to steam in circles, but returned fire with 4-inch shells which forced the U-boat to dive. Four hits damaged U-61's conning tower which

USS *Cassin* moored alongside another US Navy destroyer at Queenstown in 1918. She is painted in Dazzle type camouflage. *(NHHC)*

discouraged her commander from continuing to attack. Ingram was awarded the Medal of Honor posthumously. Another American destroyer USS *Porter* and the sloops HMS *Jessamine*, Commander Sidney Geary-Hill, and HMS *Tamarisk* arrived on the scene and protected *Cassin* throughout the night. However, no further U-boat contacts were made. The next morning, *Cassin* was towed back to Queenstown by Lieutenant Commander Ronald Niel Stuart VC in *Tamarisk*, assisted the next morning by HMS *Snowdrop*, Lieutenant Commander George Sherston. Stuart was awarded the US Distinguished Service Cross. The damaged USS *Cassin* was repaired and returned to active duty in July 1918. U-61 was sunk by HMS PC-51 a few months later.

The most successful operation by US escort vessels during the war was the capture of the U-58, Kapitänleutnant Gustav Amberger, by the destroyers USS *Fanning*, Lieutenant AS Carpender and USS *Nicholson*, Lieutenant Commander Frank D Berrien. This occurred on 17th November 1917, when an American destroyer division was escorting outward bound convoy OQ20 of eight empty ships toward its point of dispersal, with instructions to meet an incoming convoy. *Fanning's* lookout, Boatswain's Mate Daniel David Loomis, sighted a periscope in the water, close to the merchant ship, SS *Welshman*. The Officer-of-the-Deck, Lieutenant Walter Owen Henry ordered an immediate attack. The destroyer made a wide and rapid turn and depth-charged the place where it was estimated the submarine to be.

The *Nicholson* also joined the attack. Everything was quiet and it appeared that the submarine must have been missed. Then the U-58 surfaced, apparently undamaged, but stern first at a steep angle and was immediately fired upon by the guns of the destroyers. Suddenly the submarine's conning tower opened and officers and crew filed up with their arms extended shouting 'Kamarad'. The submarine had surrendered, but soon afterwards she began to sink, her sea valves having been purposely opened. The crew was rescued from the water by the American destroyers.

'The "rotund and well-fed" Amberger, wet and dripping, walked up to Carpender, clicked

USS *Fanning* (DD-37) taking prisoners aboard from the submarine U-58 which is alongside, 17 November 1917. *(NHHC)*

his heels together, saluted in the most ceremonious German fashion, and surrendered himself, his officers, and his crew. He also gave his parole for his men.' They were supplied with warm clothing, food, drink, cigarettes and, most highly appreciated, soap. Amberger later wrote to a friend, 'The Americans were much nicer and more obliging than expected.'

This was the only U-boat sunk in the war by the USN unaided. Loomis, Henry and Lieutenant Commander Hamilton Harlow, the Executive Officer of the *Nicholson* were awarded the Navy Cross. Carpender received a British medal, the DSO.

The worst reverse suffered by the US Navy at Queenstown was on 6th December 1917. The destroyer USS *Jacob Jones* was on her way back to Queenstown from special escort duty.

As she steamed independently in the vicinity of the Isles of Scilly, her watch sighted a torpedo wake about a thousand yards distant. Although the destroyer manoeuvered to escape, the high-speed torpedo struck her starboard side, rupturing her fuel oil tank.

U-58 alongside USS *Fanning* to have her crew removed after being forced to surrender, 17 November 1917. *(Library of Congress)*

US *Jacob Jones*.
(Author's Collection)

The crew worked courageously to save the ship but as the stern sank, her depth charges exploded. Realizing the situation was hopeless, Commander DW Bagley reluctantly ordered the ship abandoned. Eight minutes after being torpedoed, *Jacob Jones* sank with 64 men still on board. The 38 survivors huddled together on rafts and boats in the frigid Atlantic waters. Two of her crew, who were badly injured, were taken prisoner by U-53 commanded by Kapitänleutnant Hans Rose. Remarkably Rose also sent a message to Queenstown giving the approximate location and drift of the survivors.

Admiral Bayly was unstinting in his praise, 'We had a succession of gales during that winter of 1917–18, and I was amazed at the way all our small craft kept going, in spite of all the discomforts and continuous work. I was very proud of what they accomplished – the many acts of splendid bravery they performed and the good seamanship they displayed. Frequently I had to send a destroyer or a sloop to sea again on arrival in harbour after five days at sea; but the officers and crews were always cheerful and ready to do whatever work was at hand.' The pattern of work which they tried to achieve was the destroyers being out for five days with two days in harbour and the sloops six days out with three days in harbour.

He also greatly appreciated the work of the smaller craft and their crew, the trawlers, drifters and motor launches, manned by the RNR and RNVR. The requisitioned fishing vessels were, by the nature of their peacetime task, excellent sea boats but had none of the protective features of warships such as armour plating or watertight bulkheads. If damaged by a German torpedo, mine or shell, their fish holds could very quickly fill with water, resulting in very rapid sinking with lessened chances of survival or rescue. More than 30 were lost around their Irish patrol grounds. The events of 19th November 1917 are a good example of the hazards faced on a daily basis by the RNR Skippers and their crews. A group of eight Armed Trawlers were sweeping in pairs for mines off the coast of Co Cork, near Daunt's Rock. HMT *Morococala* struck a mine and sank in a few seconds, with the loss of all 12 crewmen, including Lieutenant Alexander Allan RNR.

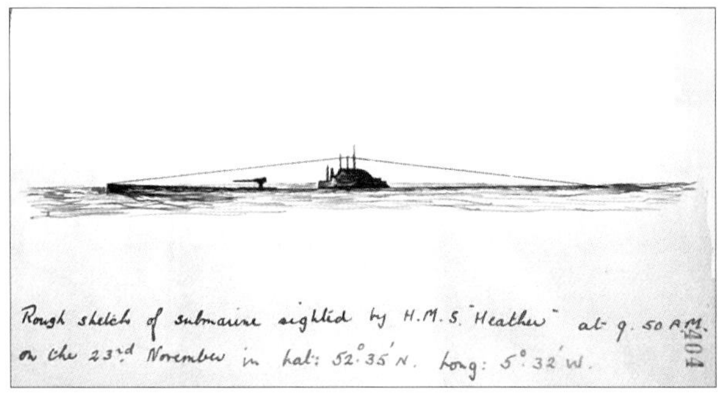

(National Archive)

Two days later, Admiral Bayly paid the following tribute, addressed to 'The Officers Commanding Armed Trawlers and Drifters on the Coast of Ireland Station:

It is with very great regret that I have learned of the loss of HM Trawler *Morococala*, blown up by a mine. During the two-and-a-half years that I have held this command I have never failed to respect and admire the devotion to duty and the seamanlike ability of the trawlers attached to the coast of Ireland. In spite of tempestuous weather, fogs, mists etc, although usually over-gunned and outranged by the submarines they have engaged with, with few comforts on board and the knowledge that they have no watertight compartments to keep them afloat: yet they are seen day after day to go out to their duties with the one idea – to destroy their country's enemies who ruthlessly prey on helpless ships, showing neither honour, manliness, nor self-respect in their cruel and brutal attacks.

We have lost some of our brother seamen from the dangers of the seas, and some from the violence of the enemy, but the same magnificent spirit continues, and I wish to express to all who serve in the Irish Trawler Force my wholehearted thanks and pride for what they have done in the past and my faith in their actions in the future.

It may be well said of our Trawler Skippers what was said of one of the first seamen in the world's history:

To tread the path of death he stood prepared. And what he greatly thought he nobly dared. [Homer's *Odyssey Book II* concerning Ulysses]

Lewis Bayly Admiral
Commander-in-Chief'

During World War I, convoys carried almost two million men to Europe. In this 1920 oil painting "A Fast Convoy" by Burnell Poole, the destroyer USS *Allen* (DD-66) is shown escorting USS *Leviathan* (SP-1326). Throughout the course of the war, *Leviathan* transported more than 98,000 troops. *(NHHC)*

Captain Knox noted:

> 'During the 18 months of war when American vessels escorted convoys through the war zone, 183 attacks were made by them upon submarines, 24 submarines were damaged and two known to have been destroyed. A total of 18,653 ships were escorted carrying vast quantities of freight to the armies in France and the civilian population of thc Allies, as well as more than 2,000,000 troops. The fact that not a single American soldier, en route to France under the protection of the United States Navy, was lost through submarine attack, is very largely due to the efficient and unremitting work of the American destroyers.'

US Submarines in action

In January 1918 seven L-Class submarines of Submarine Division Five were towed and escorted to Ireland, via the Azores. Their escorts were the submarine tender USS *Bushnell*, Captain TC Hart, who was also the CO of the Division, and the tug *Genesee*. The weather on this journey was very poor, with the submarines and escorts battered by storms and high seas. They carried out patrols from Berehaven to the southward and westward until the end of the war.

The first American submarine to complete a war patrol was USS AL10, Lieutenant James C Van de Carr, departing Berehaven on 6th March. The RN helped to train the US submariners, the first of whom to perish in the war, on 2nd March 1918, was Lieutenant Earl Childs, from the USS AL2, who was acting as an observer on board HMS H-5, Lieutenant Arthur Forbes DSO, when she was rammed by the SS *Rutherglen* in the Irish Sea in the mistaken belief

L Type Submarines Alongside USS *Bushnell* (Submarine Tender No 2) at Bantry Bay, in 1918. These submarines are, from left to right: unidentified submarine; USS L-11 (Submarine No 51); USS L-10 (Submarine No 50); USS L-1 (Submarine No 40); USS L-9 (Submarine No 49); and USS L-2 (Submarine No 41). Identification marks painted on these boats' fairwaters include the letter A, to distinguish them from British L Type submarines. *(NHHC)*

USS L-10 (SS-50) underway in Bantry Bay, during the First World War. A US battleship is in the background. *(NHHC)*

that it was a U-boat. First contact with the enemy was made on 20th March when AL4, Lieutenant Lewis Hancock, was en route from Queenstown to Berehaven. About six miles east of Fastnet, a periscope was sighted 300 yards away. The CO attempted to manoeuvre the boat into an attack positon using hydrophone bearings but lost his quarry, though not before the two vessels had momentarily collided underwater. On 22nd May Lieutenant GA Rood, commanding AL1, had a shot at a U-boat on the surface but missed, broke surface and was fired on by the U-boat, with no damge being done to either party.

Political considerations in Ireland were also a delicate matter for the Americans. On 19th May in response to the, 'Conscription Crisis, which is causing much civil and political unrest in Ireland', Vice Admiral Sims cabled the US Naval forces in Queenstown, stating the Navy Department policy that they, 'should defend themselves against actual attack, [but] they should not take part in anything which could be construed as assisting in the execution of conscription law in Ireland.'

Rosslare

The final operational base to be established under the command of Admiral Bayly was in June 1918, following the introduction of convoys between Milford Haven and Rosslare. It comprised the four Armed Drifters of the grandly named Rosslare Hunting Flotilla. They were based at the Great Western Railway ferry port in Co Wexford, which had opened in 1906. Under the command of Lieutenant L Whitehead, *Sublime II*, *Guide Me*, *Expectation* and *Sparkling Star* were ordered by Admiral Bayly to, 'patrol from dusk to dawn, carrying out drifting hydrophone watch, working with Greenore Point Hydrophone Station. In harbour to remain on short notice to proceed in daylight hours. Three drifters to be available at all times, refits and boiler cleaning to be arranged accordingly.' Whitehead reported on 9th June that a submarine was chased as far as the Tuskar Rock. Unfortunately *Guide Me* sank on 29th August, following collision off, or possibly on, The Muglins – rocks to the east of Dalkey Island.

An example of Allied co-operation

In August 1918 the airship SSZ51 out of Anglesey got into difficulties while on patrol. The report written by the Captain of the destroyer USS *Downes* is worth quoting in full:

'USS *Downes*,
16th August 1918.
From: Commanding Officer
To: Force Commander
Subject: Report of rescue of crew of SSZ51, the wrecking and salvaging of wreck

1. About 6.40 am 16th August 1918, while escorting convoy HC12, the Officer of the Deck sighted airship SSZ51 on port beam of convoy, with the mooring line trailing the water and apparently at the mercy of the wind.
2. The Officer of the Deck headed for the airship and notified me. Increased speed to 20 knots and found that the airship was making about 18 knots with the wind, which was blowing West north West, Force 3-4. I handled the ship so as to approach from windward and on coming close saw that the envelope was not full of gas and that the engine of the airship was stopped. Tried to signal the airship, but found it easier to speak by megaphone, and was requested by the pilot to take his mooring line. The airship was behaving very badly, rising and falling and making rank shears on account of the deflated envelope. My first attempt failed but in a short time I succeeded getting his line secured to the forecastle.
3. The pilot and the crew of the SSZ51 were very much concerned on account of the heat waves from the smoke pipes, fearing that the airship would catch fire, and after some conversation, requested that we send a boat out to pick them up. By 8.00 am, all the crew were safely on board.
4. I proposed towing the airship astern and informed the Senior Officer, HMS *Snowdrop*, that I would do so. The pilot feared that owing to the hole in the gas bag and the direction and the force of wind, this would not be possible, and requested that I land the car on the forecastle, and he would release the gas in the balloon and the envelope could be easily secured.
5. The airship was tending off the port bow, the car being about 40 feet away. I brought the car close to the side of the ship and threw three grapnels aboard, but owing to the extreme motion of the balloon, little progress was made.
6. A sudden puff of wind carried away one of the grapnel lines which released the airship with such a jerk that all securing lines carried away and the airship sailed high in the air and then fell slowly into the water.
7. The pilot requested me to salve as much as I could of the wreck and save the engine if possible.
8. The weight of the car when filled with water doubled the envelope in the middle, the two ends sticking up in the air. I approached the wreck slowly and picked up a

USS *Downes (NHHC)*

line trailing from the car. I hauled it to the surface and got a line through the car's bull nose. The airship was then abreast the forecastle, port side, the after end of the envelope covering the deck, and there was no means of hauling the car aboard at that place. I led a line from the car aft on the port side to the dory davits and with the authority of the pilot cut the envelope in half, stacking the after end in the forecastle and hauling the forward end aft with the car.

9. The forward end of the envelope became partly filled with water and caused a heavy strain on all lines. The car was riding bow up. The *Downes* was lying in the trough of the sea, rolling about ten degrees on a side. The dory davit was not high enough to lift the car out of the water. The after end of the car could not be reached so a strap was passed around amidships and taken to the other davit. The weight of the after end broke the car in two and apparently that part of the car and the engine sank. It seemed doubtful whether the forward part of the car could be saved, so all instruments and fittings and machine gun were taken on board. In the meantime a strong effort was made to get the remainder of the envelope aboard. Eventually the remainder of the forward part of the car was landed on deck but little progress was made with the envelope. However, the envelope had been gathered in to the side and disclosed a large mass suspended from it some 25 feet below the water. The boat anchor was used to grapple for this mass and when brought to the surface it was found to be the after end of the car and the engine. It was possible at this time to back the ship into the wind to decrease the rolling, and by use of mast whips the engine was landed on board without much further damage. After this it was not difficult to get the remainder of the envelope aboard, about 1.00 pm *Downes* proceeded to Holyhead with the wreckage.
10. During the salvage operations the pilot was seasick and unable to assist with his advice.

11. On arrival at Holyhead the wreckage, crew and all instruments saved were placed on a motor launch. I reported to the captain of the HMS *Patrol*, and received orders to oil at Holyhead join convoy OL89, Senior Officer, USS *Sterett*, at 8.00 pm, which orders were carried out.

CJ Moore, Commander USN'

Admiral Bayly was highly impressed and wrote to the Secretary of the Admiralty, drawing their Lordships' attention to, 'the very clever handling of USS *Downes* by her captain, resulting in salvaging the airship, engine, gun etc. It was a very fine piece of seamanship and remarkably well done. Commander Moore is a very useful and able officer.'

USN battleships at Berehaven

On 27th July the Chief of Naval Operations Admiral William S Benson had signalled Vice Admiral Sims, that the Navy Department, 'feels very apprehensive of at least one enemy battlecruiser getting out as a forlorn hope and attacking US Naval convoys. It is proposed to send three oil-burning dreadnoughts to Berehaven or Queenstown to be in [a] position to protect convoys, depending upon [the] British to give us information immediately of the exit of cruisers from German waters.'

In August 1918 three American battleships under Rear Admiral Thomas S Rodgers, the USS *Nevada*, USS *Oklahoma* and USS *Utah*, were dispatched to Berehaven, at Castletownbere on Bantry Bay in Co Cork, and held in readiness to engage any German battlecruisers that it was thought might venture out into the Atlantic as commerce raiders. Their arrival precipitated a revival of activity for the kite balloon section, which sent detachments and balloons to

USS *Oklahoma* (BB-37) at Berehaven in 1918. *Oklahoma* was armed with ten 14-inch guns and displaced 27,000 tons with a maximum speed of 20.5 knots. *(Author's Collection)*

USS *Utah* (BB-31) working with a kite observation balloon, in Bantry Bay, near Berehaven, Ireland, circa September/November 1918. *(NHHC)*

all three ships. On 4th October, during a severe hailstorm, lightning struck *Utah*'s balloon, which fell into the harbour in flames, fortunately with no casualties. On 14th October Rodgers received word that German warships might have escaped into the Atlantic. At the time, two troop convoys were approaching European waters. His three battleships put to sea without delay and escorted both convoys out of the potential danger zone. It was not a success for the balloonists. On the first night out lightning again struck *Utah*'s gas bag, a sudden squall also carried away *Oklahoma*'s balloon and *Nevada* lost her balloon, though it was recovered later. Luckily there were no fatalities. Otherwise the German High Seas Fleet, or any elements thereof, did not oblige or were prevented from so doing, therefore this force had no opportunity of engaging the enemy. Rear Admiral Rodgers had words of praise for other ships and their crews, 'What I saw of the handling of the two convoys was admirable and the efficiency of the British and American destroyers was admirable.' While in harbour the US sailors came ashore to the village of Rerrin, playing baseball on a diamond laid out at the Admiralty Recreation Grounds.

The USN's contribution to the mining effort has already been mentioned, laying 56,000 in total.

Submarine chasers

A squadron of 36 submarine chasers, '110 feet long gasoline-driven vessels of wooden construction, which might be regarded as enlarged motor launches' but fitted with the most

US submarine chasers. In the right background is USS SC-72. *(NHHC)*

advanced contemporary sound detection devices, sailed across the Atlantic to Queenstown in September 1918 under the command of Captain Arthur J Hepburn, 'who was most able, thoroughly reliable, never afraid of responsibility and with the divine gift of humour.' They were the first naval craft to have directional hydrophones, wireless telephones, and the new 'y gun' for firing patterns of depth charges outwards from the hull. They also had armament of a 3-inch deck gun, as well as depth charge racks astern. The Squadron was based upriver from Monkstown Bay, at Glenbrook, Passage West, with Hepburn and his staff being located at Glenbrook House. The submarine chasers made several contacts with the enemy and were credited with badly damaging one submarine, but the armistice was signed before they had really settled down to work. In all nearly 400 of these craft were constructed and 170 sent across the Atlantic to bases in Europe.

An American journalist, Ralph D Paine of the *New York Evening Post* was granted access to Admiral Bayly in 1918:

> 'I climbed the narrow, cobbled streets of that Queenstown hill and entered the Admiral's office which was cold and bare. Upon the walls were huge charts of the Irish Sea and the Western Ocean, dotted with tiny flags to indicate the positions of troop and cargo convoys and the courses of the divisions of far-flung American and British escorts. There was no other furniture than the flat-topped desk and the chair behind it, in which sat a grizzled, elderly man in a well-worn blue uniform. There was an odd illusion that the temperature of the room was falling. Soon it reached freezing point. After several years Admiral Sir Lewis Bayly raised his eyes and said in a voice from which the icicles hung, "What can I do for you, Mr Paine?"'

From this unpromising start a friendship grew, once Paine had proven himself to the Admiral. By the time he left Queenstown, the journalist regarded Bayly as, 'one of the kindliest men I have ever known.'

Captain Poinsett Pringle. *(NHHC)*

In order not to impose a burden on their host, with the exception of small quantities of fresh meat and vegetables, the US navy was self-sufficient, shipping 17 million pounds of provisions across the ocean between May 1917 and October 1918.

The Americans were under the operational command of Admiral Bayly and it is greatly to his – and their – credit that an excellent working relationship was forged, particularly with Vice Admiral William Sims, the commander of the USN's forces in the United Kingdom and Captain Joel Pringle, whom Bayly appointed as his United States Chief-of-Staff and of whom he wrote, 'Captain Poinsett Pringle would have been an outstanding officer in any navy. A thorough disciplinarian, with a well-balanced mind and a lightning brain, he was a perfect man to handle the U.S. destroyer captains, to get the most work out of them, and to keep them smiling. He was one of the greatest friends I have ever made, my beau ideal of what a naval officer should be.'

Above left: Admiral Sims' flag is raised at Queenstown. *(Author's Collection)*

Above: The First Sea Lord, Admiral Sir Rosslyn Wemyss and Captain JR Poinsett Pringle at Queenstown in 1918. *(Author's Collection)*

Chatterton admired both Bayly and Sims:

> 'Each combined with the characteristics of a man of action those of a student of naval history. Both had exhibited an outspoken attitude in regard to Departmental opinion with a freedom and courage that might have wrecked their careers, had their abilities and determination been less significant. Withal, each Admiral was embued with a deep religious instinct, the great driving force which carried them through one difficulty after another; so that, in short, either of these very modern and up-to-date Flag Officers might have been a brother to Admiral Blake or in the Cromwellian Navy.'

Their minds were so alike that Bayly relinquished command to Sims for the only leave which he took during the period of his command, 'in a US Cadillac car, driven by a US naval chauffeur, Miss Voysey and I went for a motor drive round the coast of Ireland and to visit some of my bases in June 1917.'

He wrote to Sims a little later, 'I really do think that I have the finest brand of people under me that ever man was blessed with: all I have to do is tell them to go on, and then not interfere.' The closeness and the ethos of the relationship was aptly summarised by a signal board secured to the inboard bulkhead abreast the starboard main deck gangway landing of the tender USS *Melville*. The board was placed there at the suggestion of Sir Lewis and with the complete agreement of Captain Pringle. Its message, in big brass letters, was simple, consisting of just two words, 'PULL TOGETHER.'

Chapter 12

The Final Encounters

The importance of meteorology

Gilbert Holland Price arrived at Luce Bay in 1918. He had been posted to the Meteorological Section as a junior assistant. Luckily he has left a detailed memoir to posterity. His first impression was of the 'dining hall', a canvas hangar, 'the roof and walls of which were lined with the remains of airship envelopes, as an attempt to mitigate draughts and not as a scheme of interior decoration.' He continues, 'there was no flooring and the water used in washing down the tables soaked into the earth, leaving an unpleasant smell behind it.' Nor was he impressed with the catering, 'the cooks and orderlies were the usual gang of unappetising scarecrows, and I soon found that bully beef, however unpopular generally, was regarded here as a delicacy, as it was not subject to treatment at their hands.'

The changeover from RNAS to RAF and the arrival of the DH6s had not been entirely trouble free. The new adjutant had been attempting to convert ex-naval residents to a more spit-and-polish based regime, which was resented – as was his desire to ensure that they exchanged their naval uniforms for khaki. The airship and aeroplane pilots tended not to rub along together very well either and made disobliging remarks about each other's aircraft.

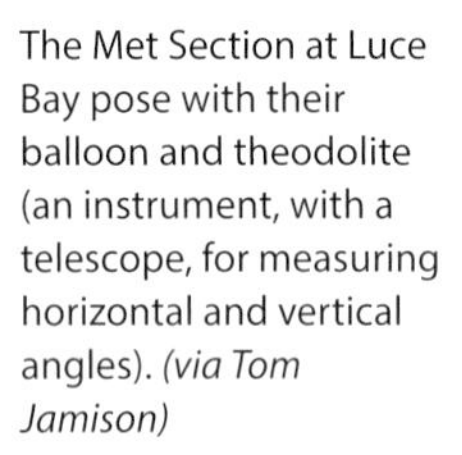

The Met Section at Luce Bay pose with their balloon and theodolite (an instrument, with a telescope, for measuring horizontal and vertical angles). *(via Tom Jamison)*

The new service was not without its teething problems. Price worked in the 'Met Hut', where a team provided a twenty-four hours a day service, supplying the raw data on barometric pressure, wind strength and direction, temperature, amount of cloud and rainfall. This was then transmitted by code from all over the UK to the Met Office in London. Synoptic charts were then drawn up and detailed local forecasts sent back for the guidance of flying officers. Additionally local wind strength at different altitudes was taken by use of small balloons which were tracked by theodolite as they ascended. At night their progress was tracked by means of a small candle suspended from the balloon with a length of twine.

Price was not particularly impressed by the social life available:

> 'On evenings when I was off duty I often went into Stranraer by the liberty lorry. There were no special attractions there, except for men who had been at Luce Bay long enough to form acquaintances among the townsfolk. It was a rather dull little seaport, which had not been improved in appearance by the siting of the gas-works – out of civic pride, perhaps? – on the waterfront. There were a small cinema, a smaller music-hall, a public reading-room, and three or four tea-shops where ham-and-egg teas could be obtained. My programme was usually the same – tea, a stroll about the sea-front, the cinema, and finally, while waiting for the return lorry, a paper of chips eaten with my fingers.'

The final days

Despite the German request on 4th October for an armistice and the fact that the U-boat campaign would be called off within 10 days, there was still time for two more atrocities. That very same day, SS *Hirano Maru*, on a voyage from Liverpool to Yokohama with 340 passengers and general cargo, was sunk by the UB-91, Kapitänleutnant Wolf-Hans Hertwig, 200 miles south of Ireland with the loss of 292 passengers and crew.

RMS *Leinster*

Then on 10th October 1918, RMS *Leinster*, a steamer operated by the City of Dublin Steam Packet Company between Kingstown and Holyhead, commanded by Captain William Birch, was torpedoed and sunk just outside Dublin Bay by the German submarine UB-123, Oberleutnant zur See Robert Ramm. It was in this ship that thousands of Queenstown officers and men had been accustomed to travel when going on leave. One of the passengers was Captain Hutch Cone USN, who was picked up unconscious after several hours in the water and ultimately recovered. Several destroyers of the Irish Sea Hunting Flotilla hastened to the scene and attempted to rescue the survivors. When the *Leinster* was first torpedoed Captain Birch ordered the ship to turn about and head back to Kingstown and ordered the lifeboats to be lowered. However, when a second torpedo struck he was blown off the bridge into the water. He

Captain Hutch Cone is the officer using crutches. *(NHHC)*

Above: RMS *Leinster* in a wartime dazzle paint scheme. *(Author's Collection)*

Above right: The B Class destroyer, HMS *Lively*. *(NMRN)*

was eventually partially pulled into the lifeboat 'Big Bertha' but was lost when the lifeboat capsized as HMS *Lively*, Lieutenant Harold Holehouse, attempted to secure it alongside.

His body was never recovered. *Lively* picked up 127 survivors, while HMS *Seal*, Lieutenant Hugh McGill, rescued 51 and HMS *Mallard,* Lieutenant Rowland Lloyd, 20, but as many as 529 died, the greatest single loss of life in the Irish Sea.

Another rescue ship was the armed yacht and former fishery protection vessel HMY *Helga*, Lieutenant EE Woodcock. Stationed in Kingstown harbour, she had shelled Dublin during the Easter Rising. She would be handed over to the Irish Free State in August 1923 and was renamed *Muirchú*, so becoming one of the first ships in the newly established Irish Coastal and Marine Service. An additional historical footnote to the story of the *Leinster* is the effect that it might have had on popular entertainment in the 1930s and 1940s. When he was seven years old the actor, singer-songwriter and comedian, George Formby, was sent to Ireland to be trained as a jockey. He was very homesick, and ran away from the stables many times. In October 1918 he planned to stow away on RMS *Leinster* and get back to England. He even sent a letter to his mother telling her that he was coming home on the *Leinster*. Luckily for him he was caught by the authorities and he missed it. He managed to get home on another boat just a few days later.

The final airship patrols

Some idea of the increased complexity of the task may be gained from this list, dated October 1918, of Confidential Books and Codes to be carried on board SSZ airships:

CB 552	Aircraft Signal Book 1917 (Aircraft edition)
CB 834	Enemy Submarine Report and Attack Table
CB 0813	Firework and Alarm Signals and Instructions for Local Signals. Home Waters
CB 0692 [9]	Air Stations W/T Code No 9
CB 0583C [2]	Standard W/T Call Signs (April 1918) (Section 2 RAF)
ID 1164	Silhouettes of Typical British Vessels (War and Merchant Vessels) October 1917

CB 707	Auxiliary Vessels Signal Book 1918
FS Form 85	Form for use in the event of forced landings
CB 532a	Cover for Aircraft Signal Books and Codes (Special Aircraft Edition)

An intelligence report for the week ending 20th October reported that on Thursday 17th October a cargo steamer, the SS *Bonvilston*, had been torpedoed off the Mull of Kintyre by UB-92, Oberleutnant zur See Johannes Paul Müller. Commodore Larne ordered SSZ11, which was already on patrol, to search the map square in which the submarine had been sighted, with SSZ12 and SSZ20 being tasked to patrol the adjacent waters.

Aeroplanes also assisted the investigation which had to be called off without success when increased wind velocity seriously impeded further activity. Further sinkings the next day saw SSZ11 sent once more to seek the U-boat which had perpetrated the action. These were the attacks on SS *Hunsdon*, which was carrying a cargo of fodder and the naval stores carrier, RFA *Industry*, Lieutenant William Norman RNR, both off the coast of Co Down, east of Ballyhornan and near the Strangford Light Buoy respectively. The latter sinking was despite having the escort of the Kingstown-based HM Trawler *Persian Empire*. Again nothing was spotted either from this airship, a convoy which was being escorted by another airship or the *Princess Maud*.

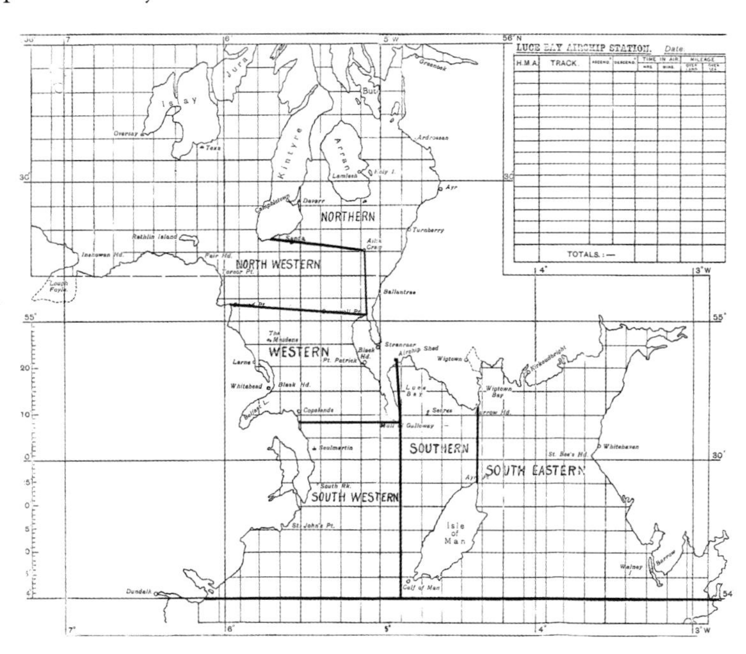

A blank daily track chart for airships patrolling from Luce Bay and Bentra. Note the named operational areas and grid squares for searches. *(Author's Collection)*

On 21st October the last merchant-vessel torpedoed in British waters during the Great War was destroyed, SS *Saint Barchan*, sunk without warning, by UB-94, Kapitänleutnant Waldemar Haumann, four miles from St John's Point, Co Down, on a voyage from Ayr to Dublin with a cargo of coal, with eight of her crew. Another sighting of a submarine on 22nd October brought considerable aerial activity, with SSZ11 dropping bombs on an oil slick at 10.00 pm. On 23rd October SSZ11 was in the vicinity of Ballykelly when heavy gunfire was reported. Neither the pilot, Lieutenant Deaker, nor the crew of SSZ12, which was escorting a convoy in the area could find anything amiss. On the same day Lieutenant Crump, flying SSZ20, escorted the *Princess Maud* for the last time before going off for a week's service afloat in HM Trawler *Corrie Roy*, which was armed with a 12-pounder.

On 3rd November a bombing attack was made by SSZ12 on a suspicious oil patch on the surface near The Maidens. Armed trawlers joined in and confirmed by hydrophone detection that a submarine was present. No confirmation of a sinking could be made. SSZ12, having been requested by Commodore Larne at 1.40 pm to patrol between Torr Head and the Maidens, had dropped her bombs at 4.05 pm and returned to base at 6.30 pm. On the

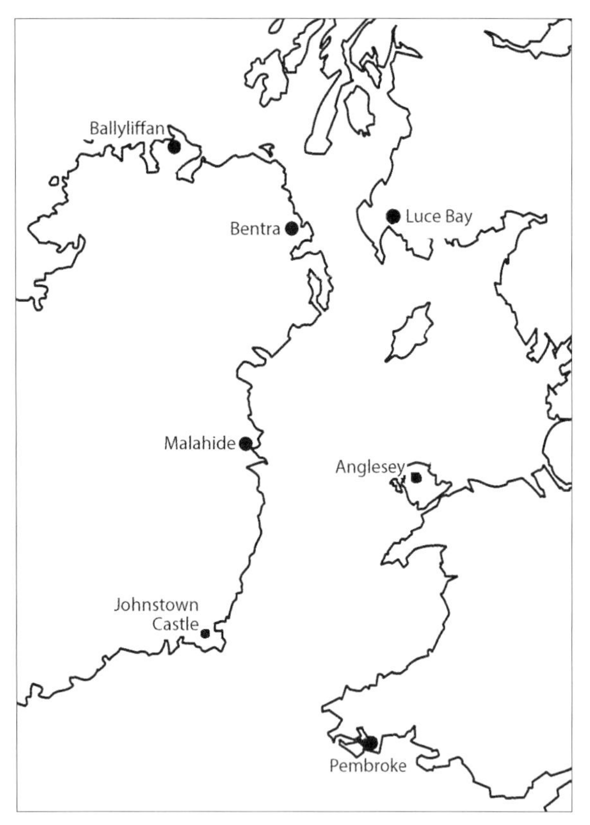

A map of the airship stations around the Irish Sea.

same day Lieutenant Knight in DH6 C2123 of 272 Squadron out of Macrihanish had to ditch while patrolling Area A (between Machrihanish, Mull, Fair Head and Inishowen Head). He was rescued by a passing drifter.

As well as Luce Bay and Bentra, the North Channel, Irish Sea and St George's Channel were guarded by RNAS Airship Stations at Anglesey and Pembroke in Wales and Mullion in Cornwall, with mooring-out sites at Ballyliffin in Co Donegal, Malahide in Co Dublin and Johnstown Castle in Co Wexford.

All this was part of a huge effort overall by the Royal Navy and the United States Navy, which combined to defeat the existential threat posed by the U-boats to the crucial trans-Atlantic traffic. Johnstown Castle and Malahide were definitely used; however, no record has, so far, been found that would prove conclusively that any airship landed at Ballyliffin. In the case of Ballyliffin it is more of a question of 'what might have been' rather than what actually took place. It was established as a mooring out station for Luce Bay and was the northernmost of British bases in Ireland. From this location airships would have provided protection for convoys rounding the north coast. The base was still under construction on 31st October 1918 and appears to have been little used before it was abandoned. Had the conflict continued, it may have been greatly expanded: it was intended to have two small steel airship sheds on site to house two Submarine Scout Twins by the summer of 1919, and the originally planned tented accommodation was due to be replaced with huts. The SST Class first flew in 1918. It used a larger 100,000 cu ft (2,800 m^3) envelope than the SS or SSZ Class types, which flew from Luce Bay and Bentra and was equipped with a streamlined and waterproofed car that could accommodate a crew of five. The envelope was 165 ft (50 m) in length, with a diameter of 35 ft 6 inches (10.8 m). It was powered by two 100 hp Sunbeam or two 75 hp Rolls-Royce engines mounted on a gantry either side of the car, and driving 9 ft (2.7 m) diameter four-bladed propellers in pusher configuration. They had better performance than the SS or SSZ Classes, a greater lift capacity, a top speed of 57 mph and an endurance of 17 hours at maximum speed or up to two days at reduced speeds. That four of the seven windbreaks on order for mooring out stations in Britain were allocated to this base gives some insight into the inclement weather conditions experienced here. Ballyliffin's location made it particularly vulnerable to gales from the north-west.

Further fixed-wing aircraft patrols were flown from the late summer of 1918 by C Flight of No 244 Squadron at Phoenix Park, Dublin and from Tallaght, North Dublin.

Malahide was first proposed as an airship station in June 1917, 'A sub-station in the vicinity of Dublin or Kingstown to be worked in conjunction and under immediate orders of Llangefni, Anglesey.' One incident of note occurred at Malahide on 1st April 1918, when SSZ50's envelope was ripped in a night-landing accident. TB Williams wrote:

A plan of Malahide Mooring-out Station. *(Cross & Cockade)*

> 'Malahide Mooring-out Station became extremely useful to us, and I often landed there and used it as my base. We were received with very mixed feelings by the local inhabitants. One man who lived very close to the gates of our castle sent me a message saying that there was a bedroom and bath always available in his house; I could just go in and help myself. On one occasion I was presented with a nice fat rabbit. My crew rigged up a little parachute on the way across the Irish Sea and we dropped it onto the lawn of my cottage. There were others who didn't like us at all.
>
> We had no telephone at Malahide at first and had to go to the local police station for messages. When returning on one occasion in the sidecar of a motor cycle, our only ground transport at the time, our speed was increased by a hail of bullets, which fortunately did us no damage. I was told that they did not like my naval uniform. Some time after that our ancient Ford, running with supplies from Dublin,

SSZ Class airship on the ground at Malahide, situated 10 miles north of Dublin. *(JM Bruce and GS Leslie Collection)*

Above: SSZ58 at Malahide in 1918. *(JM Bruce GS Leslie Collection)*

Above right: SSZ33 over Malahide sub-station. *(Author's Collection)*

> was waylaid with a rope across the road. Both the driver and the van were damaged, but not irreparably. There was a more subtle method of coercion. A row of men would sit silently on the boundary wall and just stare. It sounds nothing, but when this goes on hour after hour, day after day, it became too much for some who had to endure it. I was usually away in the air but a Petty Officer who looked after ground affairs could stand it no longer and I had to fly him back to base. I remember being in a police station in Ireland waiting for a telephone call; looking idly out of the window I saw a group of men playing pitch and toss, with some money passing. I said to the Sergeant, 'Isn't it illegal to gamble in the streets?' 'Yes, sir' answered the Sergeant. 'But while they are doing that they are not doing anything worse'.

Wexford Mooring-Out Station at Johnstown Castle was commissioned in May 1918, as a sub-station of RNAS Pembroke, with the first arrival being SSZ56 on 18th. SSZ52 joined her there on 27th, with SSZ16 replacing SSZ56 the following month. It was noted that, 'the cover is good and the ships seem to be well-protected in the cutting.'

Two SSZs were based at Wexford from 1st to 8th June, carrying out a large amount of useful patrol and convoy work. One of the crew members, H Gamble, later spoke of his time at Johnstown Castle:

> 'I think the most adventurous part of my life with the airship service was early 1918 until August 1918. We were sent across to near Wexford in SSZ56 to make ourselves

Below: A rare image of a SSZ Class airship over Co Wexford. *(Author's Collection)*

Below right: An airship of the SSZ Class coming in to land at Johnstown Castle. *(JM Bruce and GS Leslie Collection)*

a nest of our own, for in a huge park many trees had been felled and instead of a shed we tied the ship up in this avenue, the wood being its only protection. To fill the petrol tanks I had to scale the trees on either side while the landing party pulled the ship over to me. Unable to get much gas sent over to us we had to wait until the sun shone upon the envelope to swell the gas and give us lift before we could get out on patrol. In the cool of the evening, of course, when we returned to land, we would be very heavy and landing was difficult. To assist us on our journey home one night, three German submarines gave us a good shelling. The cheeky bounders were only about two or three miles from Tuskar Rock and the lighthouse keeper watched them, needless to say we did not go after them because we were such a good target.'

On another occasion his airship co-operated with an American destroyer off Waterford to give a U-boat a thorough pasting with bombs and depth charges.

On 9th June SSZ56 had to be deflated as its gasbag was ripped during a storm, the parts were returned by tramp steamer to Pembroke. No airship was on station thereafter until 26th June when SSZ16 arrived for duty. During its time there, the pilot of SSZ16 had to request assistance from the USNAS at Ferrybank, Wexford, in attempting to land in high wind. Personnel were sent over without delay and helped to moor the airship.

It is also of interest to note that Pembroke had at least one US airship pilot, Ensign Norris USNRF, who had to make an emergency landing in the Irish Sea in SSZ37.

Having repaired a leak in the water circulation system, he was able to resume his flight and return to base. This airship was based at Wexford in July.

USNAS Ferrybank, Wexford – note the washing drying on the hedgerow. *(Delaney Collection)*

Above: Posing with a camera from the cockpit of an airship. *(Author's Collection)*

Above right: SSZ37 over a convoy in October 1918 in a photograph taken from SSZ16. *(National Archive)*

Later reports noted that a W/T station had been set up, communication by telephone laid on, a bomb dump and a type-A silicol plant for the manufacture of hydrogen created, as well as an Officers' Mess and a canteen for the men. Flying from Wexford would – in all probability – have ceased soon after the Armistice.

Before the war ended plans had been made to introduce a much larger type of airship, the rigid, Zeppelin-type 33-Class, to patrol out into the Atlantic. Preliminary work began on a suitable shed near Lough Neagh (it was originally proposed in 1916 as a major rigid airship construction site, effectively a Naval Airship Dockyard immune from enemy air attacks and further earmarked in 1917 as a base for rigid airships) but the Armistice was signed before this was completed. Another similar station was under construction at Ballyquirke, near Killeagh in Co Cork, 20 miles to the east of Cork City. When work was discontinued, the structures on site consisted of one rigid airship shed and one Coastal Class airship shed, both partially completed, as well as a significant number of ancillary buildings.

There were also plans afoot to replace the DH6s with larger and much more capable twin-engine Vickers Vimys, a type made famous in June 1919 by being flown across the Atlantic Ocean from St John's Newfoundland to Clifden in Co Galway by John Alcock and Arthur Whitten Brown.

On Monday 11th November 1918 the Armistice was signed and the 'war to end all wars'

Below: HM Submarine C.17 photographed from a height of 50 feet by the crew of SSZ16 in October 1918. *(National Archive)*

Below right: Kil Class Patrol Gunboat, photographed from an airship. *(National Archive)*

came to its conclusion. Gilbert Holland Price described the scene at Luce Bay:

> 'Rumours of an end to the fighting had been abroad for a day or two when the news of the signing of the Armistice came through on the morning of November 11th. At eleven o'clock we heard sounds of revelry. Soon the men were marching in procession round the station, headed by an impromptu band consisting mostly of tin cans, and an equally impromptu drum-major twirling a broom. He was accompanied by somebody trundling a wheelbarrow. Dinner was graced by a visit from the orderly officer, supported – literally, for he was very drunk – by a colleague.'

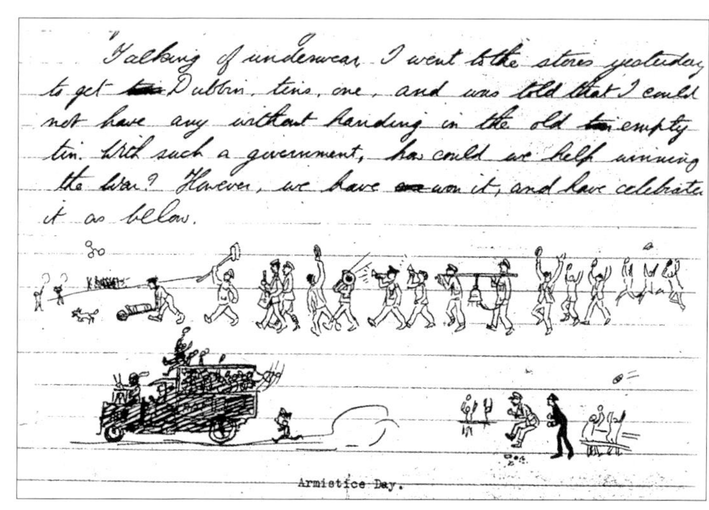

Talking of underwear I went to the stores yesterday to get Dubbin, tins, one, and was told that I could not have any without handing in the old empty tin. With such a government, how could we help winning the War? However, we have won it, and have celebrated it as below.

Armistice Day.

Gilbert Holland Price's sketch of the 'parade' on 11th November 1918. *(Author's Collection)*

In Larne the news was made known to the people of the town, as reported in the *Larne Weekly Times*, 'by courtesy of a communication to the chairman of the Urban Council from Commodore Carpendale CB. Though not unexpected the magnitude of the news had at first a bewildering effect and its meaning could hardly be realised. The full-throated announcement from every ship in the harbour soon banished all doubts and uncertainties, and at once the residents gave themselves up to rejoicings.' Workers from the factories, foundaries and the shipyard paraded through the town singing patriotic songs amidst a profusion of flags and streamers hung from shops and homes, with hundreds of children out on the streets waving small Union Jacks. The festivities were quickened by the arrival of the

Mine sweeping trawlers in Larne Lough. *(Larne Museum & Arts Centre)*

Above: The Olderfleet Hotel, Larne, which was the location of the Royal Naval HQ in the First World War. *(Larne Museum & Arts Centre)*

Above right: Drifters being maintained by the Larne Shipbuilding Company. *(Larne Museum & Arts Centre)*

Naval Brass Band early in the afternoon, followed soon by other local bands. The departure of the *Princess Maud*, under the command of Captain Hamilton, was specially hailed, saluted by every ship in the harbour, to which her own siren and gun responded. 'Everyone felt that the officers and crew from captains M'Calmont and Hamilton to the youngest boy had day after day proved themselves to be real heroes, making a dangerous crossing twice daily almost without intermission, since war was declared.' Then, 'For about an hour that evening huge crowds visited the harbour, where from the naval vessels, brilliantly illuminated, was given a splendid fireworks and searchlight display.'

Captain Gordon Campbell recalled in his autobiography an incident not long after the Armistice which also involved Tom Elmhirst. Major Elmhirst was the CO at Llangenfi (RNAS Anglesey), his counterpart with No 244 Squadron at Bangor (RAF Aber) was Major Harold Probyn, who boasted of flying his DH6 under the Menai Bridge. Elmhirst could

Larne Naval Base Brass Band. *(Larne Museum & Arts Centre)*

not resist the challenge that he could not do the same in 'one of his rotten airships.' He came to Campbell for permission and with a plan showing four feet to spare, two above and two below. Campbell agreed with the proviso that he should accompany Elmhirst, 'We flew clean under (in SSZ73), I think I was surprised as anyone when we got through. I quite expected a rap over the knuckles but at that time the Air Force and the Navy were somewhat mixed up, and perhaps no one knew who was responsible.' It was, without doubt, a quite remarkable feat of airmanship.

Two destoyers, USS *Stevens*, Commander R Williams, and USS *Duncan* paid a courtesy call to Belfast, mooring at the Alexandra Jetty between 16th and 19th November, under the overall command of Commander Rufus F Zogbaum Jr of *Stevens*. Soon they would be departing Queenstown, with 23 destroyers on their way back home in December.

USS *Stevens (NHHC)*

USS *Duncan (NHHC)*

Chapter 13

Other U-boats destroyed around the coast of Ireland between 1916 and 1918

To complete the record, the following naval encounters around the Irish coast not already described in the text above, resulted in the destruction of these U-boats.

U-68, Kapitänleutnant Ludwig Güntzel, was the first kill made by the Q-ship HMS *Farnborough*, Lieutenant Commander Gordon Campbell, on 22nd March 1916 in the Irish SW approaches, off Dunmore Head in Co Kerry, using depth-charges and gunfire. Campbell later wrote:

> 'To the men concealed at the guns and elsewhere this was the first great test of the discipline and drill we had been training for, as it was obvious that the submarine might fire another torpedo and perhaps successfully. All remained quiet, and the men, lounging about, continued to smoke their pipes. One young seaman was whistling at his gun, because, as he explained when asked what he was doing, "if he didn't whistle he would get scared".'

Campbell was promoted to Commander and awarded the DSO.

U-83, Kapitänleutnant Bruno Hoppe, was sunk by gunfire from HMS *Farnborough*, Commander Gordon Campbell, SW of Ireland on 17th February 1917. *Farnborough* was torpedoed by U-83 and a 'panic party' scrambled to the lifeboats. When the U-boat surfaced to finish off its kill, the White Ensign was broken out and the hidden guns and their crew were revealed. U-83 was destroyed but *Farnborough* was in a scarcely better state and had to be towed into Berehaven by the sloops HMS *Buttercup* and HMS *Laburnum*, Lieutenant Commander William Hallwright. Gordon Campbell was awarded the Victoria Cross and the gallant *Farnborough* was scrapped.

Laburnum's other claim to fame was that in April 1916, during the Easter Rising she shelled the outskirts of Galway, the ship's log stating, 'Hands to Action Stations. Fired nine rounds from the after gun, give common and four lyddite in direction of rebels advancing on Galway town.' The Senior Naval Officer in Galway, Commander Francis Hanan, was an Irishman. An illuminating story about his time serving in Galway was told by a New Zealander, Sub Lieutenant Oscar Carter. He was involved in patrols against Sinn Féin as First Lieutenant of ML 374. It was thought that the Germans were landing spies, as well as arms and agitators on the West Coast of Ireland. Carter describes it as an unpleasant area with strong westerly winds and heavy seas, but with a number of places affording good shelter. They got to know many of the small towns such as Kilrush, at the mouth of the River Shannon where they had good relations with the locals. The bank manager even invited the crew of the ML for Christmas dinner. Neither he nor his wife was at home, but a sumptuous dinner was laid out, including Irish whisky and stout. Having enjoyed themselves immensely, they left a note of thanks, a large bag of brown sugar (a rare luxury at the time) and a tin of tobacco. They later learned that the bank manager was the head of the local Sinn Féin and highlights that there was no antipathy towards the RN by Sinn Féin, just a hatred of the British government.

Still on the subject of MLs, Captain Bill Dykes of No 105 Squadron, which had a flight based at Castlebar, ostensibly to hunt for submarines but in reality on internal security duty, later recalled:

> 'Our Navy kept a few small Motor Launches (ML) in the various west coast bays. I spotted an ML lying at anchor just off Westport in Co Mayo when I was carrying out a solo patrol there in my own RE8, C2368; I observed that this ML was displaying on her deck the wrong code sign for the day. Was it friend or foe? I drew my observer's

A flight of RE8s from No 105 Squadron was based at Castlebar. *(via Philip Archer)*

attention to this fact and we decided to draw the skipper's notice to his error by the only means of communication available to us. We had no radio so I wrote a message, "Correct your code sign for the day" and placed this in a message bag. I then flew low over the boat, just clearing the mast, and more by luck than good judgement managed to land the message bag on his deck. My manoeuvre brought almost the entire crew on deck. I climbed to three or four hundred feet circling around to await results. The code sign was promptly rectified. I then flew low alongside and waggled my wings, receiving friendly waves from all on deck. More obvious British tars one could not imagine. I was glad I had not taken hostile action and loosed off my bombs. I had probably caught them at lunch time and it would have disturbed their pleasant afternoon snooze on a nice calm and very hot day.'

The Senior Naval Officer, Galway, was less than impressed by the efficiency of some of the MLs' RNVR officers, writing:

'Although there are some conspicuous exceptions, and also a considerable improvement in general, yet, the majority of the Officers in the MLs do not get the maximum efficiency out of their vessels. They are, as a class, men who have lived sedentary and easy lives and lack the staying power necessary for persistent sea keeping under the extremely hard conditions experienced in MLs. (There has been more sickness amongst them than amongst any other class in this area.) It is submitted that a trial might be made of a few groups of MLs officered by Midshipmen and Acting Sub Lieutenants RN and POs RN as coxwains, with divisions under a Lieutenant or Sub Lieutenant RN. It is recognised that the technical training of Midshipmen might suffer, to some degree, but the experience in command should at least counteract this, and, even if it did not, in view of the undoubted fact that the whole result of this war hangs on the success or failure of the anti-submarine campaign, this loss in the future counts as little in the face of the current necessity.'

ML189 on patrol. *(via Commander Rob Milligan)*

The battleship HMS *Albemarle*. *(NMRN)*

In other words the work of the MLs was, in his opinion, of greater use to winning the war than swinging around a buoy in a battleship at Scapa Flow.

The RN and the Easter Rising 1916

In command at Berehaven was another Irishman, Commodore Hugh Heard, who owned property not far away. Admiral Bayly had received temporary reinforcements at Queenstown, the battleship HMS *Albemarle*, Captain Raymond Nugent, the cruiser HMS *Gloucester*, Captain William Blunt, and 2000 Royal Marines in TSS *Great Southern*.

One of the 'booties' was stationed as a sentry at Admiralty House; one foggy night a mysterious, silent figure failed to answer his challenge. He fired a single shot and the intruder perished instantly. On examination it turned out to be an entirely innocent donkey rather than an IRA assassin. The Admiral noted that the incident cost him £2, one to the sentry for his markmanship and another to the old lady who owned the donkey. *Albemarle* was stationed in Ringaskiddy Bay and a rumour was put about that Cork would be shelled with her 12-inch guns if there was any sign of trouble, *Gloucester* was dispatched to Galway with *Laburnum* and *Snowdrop*, while the Admiral's flagship, the Scout Cruiser, HMS *Adventure*, Captain George Hyde, went to Kingstown to act as a wireless relay station in the event of any telephone lines connecting Dublin to the outside world being cut.

The sloops HMS *Zinnia*, Lieutenant Commander GFW Wilson, and HMS *Bluebell*, Lieutenant MAF Hood, intercepted the 'Norwegian' steamer '*Aud*' off the mouth of the River Shannon. On arrival off Queenstown, watched from the veranda of Admiralty House by Admiral Bayly and Miss Voysey, 'the *Aud* stopped, hoisted two German naval ensigns, and lowered her boats, into which got about 30 officers and seamen in naval uniform. Then an explosion occurred and she went to the bottom.'

The crew had scuttled her, the ship was found to be the German auxiliary, *Libau*, Oberleutnant zur See Karl Spindler, with a large cargo of arms and munitions for the rebels. Rebel action in the southwest was also somewhat handicapped by the fact that the IRA cut the telegraph wires before the anticipated signal for the rebellion to start was sent. A considerable detachment of troops was sent to reinforce the guard at Skerries W/T station. One further Royal Navy action is worthy of note.

The following letter, dated 3rd May, 1916, is addressed to the Senior Naval Officer, Belfast, by Lieutenant Colin MacLeod Campbell RN, who was Commanding Officer of the destroyer HMS *Tigress*, concerning an operation conducted by an armoured train under his command:

> Sir,
> I have the honour to report the movements of the armoured train under my command on the 30th April and the 1st May 1916. In accordance with orders received from GOC, Belfast, I left Belfast at 6.00 am on Sunday, April 30th, in charge of an armoured train, which consisted of an engine, two trucks and a van. The Engine and Trucks were armoured with 3/16ths plating, and the latter were fitted with loopholes; a maxim gun was mounted in each truck. The personnel consisted of three Officers and ten men RN, two Officers and 64 NCOs and men of the 10th Royal Irish Fusiliers.
>
> Train arrived at Dundalk at 7.45 am, and I reported to the OC there, who informed me that everything was quiet in his district, and that he had cyclist patrols out in every direction. The train left Dundalk at 8.45 am, and steamed slowly down the line, stopping at Fane, Castlebellingham and Dee bridges, which were guarded by the Military. The officer in charge at each Bridge reported everything quiet.
>
> The train was stopped at Dunleer in order to gain information of rebels, reported to be in Barmeath Castle, just south of Dunleer. I ascertained from a Sergeant in the RIC, who had personally visited the Castle, that the rebels had left two days previously, and had not returned. The train arrived at Drogheda at 10.30 am, and I proceeded to see the OC, to arrange a plan of utility with him. I was ordered to proceed to Oldcastle, to visit the German internment camp, and to stop at Navan and Kells on the way. I saw the OC Oldcastle, who informed me that he had 600 prisoners with 150 men to guard them. The prisoners were giving very little trouble, and everything was quiet in his district.
>
> I arrived back at Drogheda at 4.15 pm and left for Amiens Street Station, Dublin. At 5.30 pm I stopped at Rush and Lusk and ascertained from the Officer in Charge that 30 rebels in his district had surrendered during the course of the day. I then proceeded to Donabate (the up line between Donabate and Rush and Lusk was out of action owing to part of a bridge having been blown up by rebels; repairs were in

A rare image of the SS class airship over Dublin in 1916. *(Author's Collection)*

progress, and I was informed that the line would shortly be open to traffic).

I arrived at Amiens Street Station at 7.30 pm, and reported to Headquarters, who ordered me to report again in the morning. I proceeded to Headquarters on Monday forenoon, 1st May, and was ordered to return to Belfast with the armoured train. Everything seemed to be quiet in the Amiens Street area, except for occasional sniping. The train left Amiens Street Station at noon, and arrived in Belfast at 5 pm. I observed everything in order on the line on the way back.

I am strongly of the opinion that the presence of the armoured train at different places along the line exercised a very good moral influence upon the people in the districts through which it passed.

I would like to bring the name of Mate (E) Allen to your notice, for the efficient manner in which he designed and constructed a shield for one of the maxim guns. I should further like to remark upon the very good comradeship which existed between all ranks and ratings of the Army and Navy under my command.

In the aftermath of the Rising, Dublin was also visited by an airship from RNAS Pembroke, to 'show the flag' and to take photographs of the destruction caused to many buildings in the city.

Returning to the subject of submarine sinkings, U-81, Kapitänleutnant Raimund Weisbach, which had just torpedoed and sunk the Tanker SS *San Urbano* was itself torpedoed in the Atlantic to the west of Ireland on 1st May 1917 by the Irish-based, British submarine E-54, Lieutenant Commander Robert Raikes. The captain of U-81 and seven of his crew were saved. Weisbach was injured and was rescued by one of the British submarine's officers who jumped into the sea with a line. Admiral Bayly wrote, 'It was a wonderful feat, very skilfully and humanely carried out. He arrived at Queenstown about 1.00 am and came up to see me at Admiralty House. We sat on the kitchen table, while I listened to his story and my steward made us some coffee.' The C-in-C puffed on his pipe while Raikes smoked cigarettes that belonged to the Admiral's valet.

UC-29, Kapitänleutnant Ernst Rosenow, was laying mines off Valentia Island on 7th June 1917, when it encountered and torpedoed the Q-ship, HMS *Pargust*, Commander Gordon Campbell, to which he had transferred with his crew from *Farnborough*. Similar tactics were successful, UC-29 was sunk, and though badly damaged, *Pargust* lived to fight another day. She was attended by the USS *Cushing*, Commander David C Hanrahan and HMS *Zinnia*, being taken in tow to Queenstown by the sloop HMS *Crocus*, Lieutenant Commander George Skinner. Campbell was promoted to Captain and received a bar to his DSO. Lieutenant Ronald Niel Stuart and Seaman William Williams were both awarded the Victoria Cross by ballot of the crew. Admiral Bayly praised his coal-burning sloops and their COs thus, 'they were excellent sea boats and the youngsters who commanded them were splendid.' While hardy vessels, the fact that the sloops burned coal rather than oil was a limiting factor, as they could not be kept going like oil-burning modern destroyers, it took them one day of their time off in harbour to coal, and in bad weather longer. Bayly went on to add:

> 'Looking back on the two years of my command of the Western Approaches [before the US Navy took a part], when the whole of our seagoing trade was guarded, escorted, or rescued by sloops, trawlers, motor-boats, and mystery ships, all of them commanded by young naval, Naval Reserve, Naval Volunteer Reserve, and trawler skippers, in all weathers, off a stormy coast, with as little time in harbour and as much time at sea as was humanly possible, I feel that the country owes them a great debt of gratitude. I can truly say that they never failed me, and no complaints reached me; there was always a readiness to do not only what was possible, but also sometimes what appeared to be impossible. No one in command of a squadron could be more proud of it than I was – and am – of my brave little heterogeneous fleet and my hard-worked staff.'

Rear Admiral Francis Miller. *(Author's Collection)*

He continued, 'Rear Admiral Miller at Lough Swilly and Captain Carpendale at Larne ran the north side of the Irish Sea so well that I was able to devote myself to the main trade routes to the south.' Miller himself, who had hoisted his flag at Buncrana on 18th July 1917 (succeeding retired Admiral, Captain Frank Finnis, who had been in post for a year) had high praise for the commanding officer of the 2nd Destroyer Flotilla, Captain John Sparks CBE, writing:

> 'He has been in command of the 2nd Destroyer Flotilla under my orders from September, 1917–December, 1918. I desire to recommend this officer most

> strongly. He has worked zealously, and displayed pronounced ability in securing good organisation and discipline, thus maintaining the flotilla in a high state of efficiency. This, together with careful training and instruction in both offensive and defensive measures has enabled the flotilla as a whole to secure protection for the convoys and to meet with a fair measure of success in hunting, destroying and damaging enemy submarines.'

Miller also singled out two other officers, whom he recommended for military OBEs, firstly, Lieutenant Commander Kenneth Swan RNVR, writing:

> 'Ever since August, 1914, this officer, in addition to the duties for which nominally borne, has voluntarily undertaken any special duties which were required to be carried out, both whilst in HMS *Cyclops* [at Scapa Flow] and at Buncrana. In private life he is a well known barrister and since he joined in 1914, he used his great abilities, and his unflagging energy, for the good of HM Service, and has taken a prominent share in the organisation of the Bases at Scapa Flow and Buncrana. He has ever had the welfare of the men at heart and has been indefatigable in his efforts to arrange for their amusements and recreations, especially during the long winter nights in the far north. I should think this officer has put in an average of nearly 16 hours a day, and has carried out arduous and responsible duties in an exceptionally efficient and capable manner.'

Secondly, Lieutenant Commander Reginald Bates:

> 'The zeal, energy and ability of this officer have contributed largely to whatever success I may have attained in the creation, development, co-ordination and general organisation of two new and highly important war bases – the Grand Fleet Base at Scapa Flow, and the Convoy Base at Buncrana. At both Bases we started with inadequate and untrained staffs, and from May, 1917 [the date of starting this Base], until August, 1918, Mr Bates was my principal assistant in all operations branch [Convoys, Escorts, Intelligence, Communications etc.] in addition to his secretarial and administrative duties. He kept a night on duty for operations alternately with myself and my Chief of Staff, referring to me only in cases of exceptional difficulty. These duties involved the safety and welfare of many convoys, and have been exceptionally responsible and arduous [particularly for a secretary] and have been carried out in an exceptionally efficient and capable manner. Throughout he has displayed unflagging zeal, energy and devotion to duty, sparing himself, neither by day or night. In August last, with the reorganisation of the base staffs by the army, the duties previously performed by Mr Bates at Buncrana were divided up amongst a large number of officers.'

Buncrana's importance increased with the introduction of the convoy system and the basing there of half a dozen sloops from Queenstown in the summer of 1917.

UC-44, Kapitänleutnant Kurt Tebbenjohanns, struck a German-laid mine off the coast of Co Waterford on 4th August 1917. At 10.20 pm a loud explosion had been heard by the inhabitants of the fishing village of Dunmore East. When rescued from the sea by two locals in a rowing boat, the captain was most indignant and complained that the British had been very lax in not sweeping the area properly!

He was actually the victim of a clever plot hatched by Admiral Bayly and by the Admiralty code-breakers in Room 40, which lured the Germans into thinking a previously laid minefield had been swept, when it had in fact been left to trap the unsuspecting U-boat. Much valuable intelligence material was found when the boat was raised and examined, in which process HMS *Snowdrop* participated.

UC-42, Oberleutnant zur See Hans Albrecht Müller, was lost in the explosion of her own mines off Cork on 10th September 1917.

UC-33, Oberleutnant Alfred Arnold, was shelled then rammed by HMS PC-61, Lieutenant Commander Frank Worsley RNR, (who had been the captain of Ernest Shackleton's SS *Endurance* in Antarctica from 1914–16) in St George's Channel on 26th September 1917. The PC Class were built with a mercantile superstructure and operated as Q-ships.

U-87, Kapitänleutnant Freiherr Rudolf von Speth-Schülzburg, was rammed by HMS *Buttercup*, Lieutenant Commander Arthur Petherick, an Arabis Class sloop, and depth-charged to be finally sunk by HMS PC-56, Lieutenant William Florence RNR, in the Irish Sea on 25th December 1917. The U-boat had just sunk the SS *Agberi* which was built at Belfast in 1905 by Workman, Clark & Co Ltd for the Elder Dempster Line. She was en-route from Dakar to Liverpool carrying passengers and a cargo of silver and ivory. Both *Buttercup*

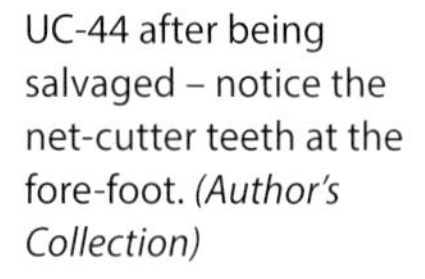
UC-44 after being salvaged – notice the net-cutter teeth at the fore-foot. *(Author's Collection)*

and PC-56 made frequent use of the facilities at Kingstown.

The moment a cargo ship is torpedoed by a U-boat *(Author's Collection)*

U-110, Korvettenkapitän Carl Albrecht Kroll, was sunk on 15th March 1918 north-west of Malin Head. She was found and depth-charged by the destroyers HMS *Michael*, Lieutenant Commander Alfred Englefield Evans, and HMS *Moresby*, Lieutenant Commander Godfrey Chambers. U-110 had been caught submerged at 130 feet after torpedoing the SS *Amazon* off Malin Head. Six depth charges shook the submarine badly and her diving-gear was seriously damaged by the explosions. Seeking safety in depth Korvettenkapitän Kroll took her down to 300 feet but the increasing pressure caused numerous leaks in the hull and he was forced to bring the U-boat to the surface to avoid disaster. *Michael* was five miles away when the submarine appeared but she opened fire immediately and Leutnant zur See Busch was killed as he led the gun's crew out on deck for a futile last stand. More hits were scored and Kroll, assembling his men on deck in their life-jackets, ordered them to jump into the sea. U-110 was by now half awash after being rammed by one of the destroyers. She sank rapidly leaving only ten survivors swimming in the water. Kroll, and the remainder of the crew, went down with his boat. The survivors were found to be very young and inexperienced, as evidence that Germany's resource of manpower was dwindling. Lieutenant Cyril Bower of *Michael* was awarded the DSC.

HMS *Kilfree*, a 'double-ended' patrol vessel – so designed to confuse the enemy. *(Author's Collection)*

U-61, Kapitänleutnant Victor Dieckmann, was sunk in a depth-charge attack in St George's Channel by HMS PC-51 on 26th March 1918. The original intention had been to ram until the helm jammed at the last moment.

U-104, Kapitänleutnant Kurt Bernis, was engaged by the USS *Cushing*, Commander William D Puleston, on 23rd April and subsequently depth-charged and sunk by the sloop HMS *Jessamine*, Commander Sidney Geary-Hill, in St George's Channel on 25th April 1918. Geary-Hill's report to Admiral Bayly noted that the U-boat had been spotted on the surface at 1.45 am. He ordered full speed ahead, saw the submarine diving just in time and dropped four depth-charges. It then rose out of the water, 'split open.' A survivor, Engine Room Artificer Karl Eschenberg, told his captors that the U-boat had made two unsuccessful attempts to sink Holyhead to Kingstown mail steamers.

U-103, Kapitänleutnant Claus Rücker – one of the regular duties for Queenstown's US destroyers and RN sloops was escorting the great transatlantic liners, in use as troopships.

The weekly reports make repeated mentions of the *Aquitania*, *Olympic* and *Mauretania*. In the early hours of 12th May 1918, while en route for France with US troops under the command of Captain Bertram Hayes, a surfaced U-boat was sighted from *Olympic* 1600 feet ahead. *Olympic's* gunners opened fire, the ship turned to ram, striking U-103 just aft of her conning tower with her port propeller slicing through the U-boat's pressure hull. *Olympic* did not stop to pick up survivors, but continued on to Cherbourg. Meanwhile, the USS *Davis*, Lieutenant Commander WV Tomb, one of four US destroyers in the escort, had sighted a distress flare and picked up 31 of the crew. It was later discovered that U-103 had been preparing to torpedo *Olympic* when she was sighted, but the crew were not able to flood the two stern torpedo tubes. Captain Hayes was awarded the DSO. A commemorative plaque was later presented by the 59th Regiment United States Infantry and was displayed in one of the *Olympic*'s lounges.

UB-65, Kapitänleutnant Martin Schelle, was lost in a mysterious explosion on 10th July 1918. The submarine USS AL2, Lieutenant Foster, was working out of Berehaven and spotted the U-boat's periscope but before it could attack the U-boat blew up. Two theories have been considered, firstly that the U-boat was blown up by the premature detonation of a magnetic pistol in the nose of one of its torpedoes, or alternatively a second U-boat, known to be in the area at the time, destroyed UB-65 by mistake.

Chapter 14

Some Final Thoughts

By December the airships had ceased to fly and Bentra returned to its rural calm and peaceful obscurity. The farmland was reclaimed and another farmer, Mr Service, was able to make use of the buildings left behind. Some of the huts were used as holiday homes and later as permanent homes until the late 1950s. The sinkings continued into 1919 as ships hit drifting but still deadly mines. As for the *Princess Maud*, she served on until 1932 when she was scrapped following a grounding. Commodore Carpendale was succeeded in Larne as SNO by a former CO of HMS *Garry*, Commander Werden Wilson, for a few months before the naval base was closed.

The Coast of Ireland Station was much reduced with Commander Francis Crean at Killybegs and Commander CR Sharp at Berehaven. Tom Elmhirst would achieve high rank in the RAF in the Second World War as also would John Cole-Hamilton. Elmhirst was Second-in-Command of British Air Forces in North-West Europe on D-Day; later he was the first Commander-in-Chief of the Indian Air Force and finally he served as the Lieutenant-Governer of Guernsey. He died in 1982 at the age of 86.

In November 1941 Cole-Hamilton returned to Northern Ireland as Air Officer Commanding in the rank of Air Vice Marshal. He died in August 1945 at the age of only 50, when AOC of No 11 (Fighter) Group. Irving Hartford lived to a ripe old age, it is believed that he was about 90 years old when he passed away in 1980. George Colmore attained the rank of Squadron Commander and died in 1937 at the age of 52. Thomas Philip Yorke-

A captured German U-Boat, now flying the White Ensign over the Imperial German Navy Ensign, in Larne Lough. *(Larne Museum & Arts Centre)*

Moore (1895–1985) was born in Jamaica. After serving at Luce Bay, he flew in Coastal Class airships from Mullion in Cornwall, becoming one of the world's most experienced airship pilots, with over 2000 hours flying. He attained the rank of Squadron Leader in the RAF. Bert Crump settled in the West Midlands as a company director, Scout Commissioner, Freemason and magistrate. He too was relatively young when he passed away in 1956 at the age of 57. Squadron Leader Ernest Johnston became the navigator of the R.101 and died in the wreck of that ill-fated airship on 5th October 1930. Bernard Hemsley later served in the Royal Irish Constabulary, the Royal Ulster Constabulary, the RAF for a further 20 years and eventually as Assistant Commissioner of Prisons in Kenya, being awarded an OBE to add to an MBE (Military). Welshman Edward Turnour passed away in 1968 at the age of 72. Vice Admiral Charles Carpendale hauled down his flag in 1923 and for the next 15 years was Deputy Director General of the BBC, being raised from CB to KCB in 1932. During the Second World War he worked at the Ministry of Information and then as the honorary librarian for the Royal College of Surgeons. Later, being a skilled craftsman he undertook, without charge, the rebinding of hundreds of books for the library. He was masterful but a splendid mixer and endeared himself in each of his careers by his cheerful good sense, simple friendliness and absolute integrity.' He was in his 94th year when he died in 1968.

Sir Charles D Carpendale 1938 on retirement from BBC. *(Author's Collection)*

Two RN submarines moored alongside a depot ship at Kingstown. *(National Archive)*

U-19 was surrendered on 11th November 1918, and was later broken up for scrap. Its other claim to fame was as the vessel which carried Sir Roger Casement to Banna Strand in Tralee Bay on Good Friday, 21st April 1916. Its main gun was donated to the people of Bangor, Co Down and today sits near the War Memorial in the town's Ward Park. It was donated by the Admiralty in recognition of the valorious conduct of a native of Bangor, Commander The Hon Edward Bingham, who was awarded the Victoria Cross for his bravery in the Battle of Jutland in July 1916, where he commanded the destroyer HMS *Nestor*.

In April 1919 Admiral Bayly was succeeded at Queenstown by Admiral Sir Reginald Tupper for a period of two years, in which Ireland would suffer considerable bloodshed. Tupper later wrote, 'It was a very difficult time, but the events of that period, and the breaking away of Ireland after I had left, are so painful to look back upon that I have never felt disposed to talk about them.'

Just as Admiral Bayly greatly appreciated the USN's contribution, the feeling was mutual.

After the war the US naval officers who had served under his flag at Queenstown formed themselves into a Queenstown Association, with a membership of more than 500. The Admiral and his niece, Miss Violet Voysey, 'a charming young woman (I shall guess to be about 30) keeps house for him' visited the USA twice in 1921 and 1934 as the guests of

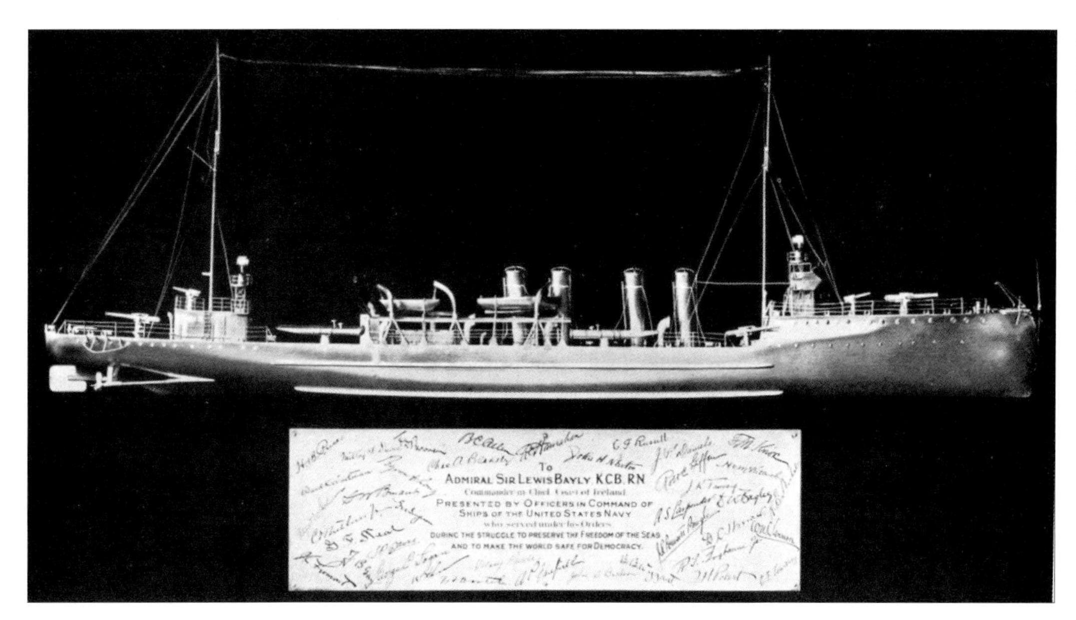

A model of one of the US destroyers that was presented to Admiral Bayly by the American commanding officers at Queenstown. *(Author's Collection)*

the Association and were 'not allowed to buy even a postage stamp' so generous was the hospitality given. Admiral Sims paid tribute to Miss Voysey in 1921:

> 'Although our 'Old Man' at Queenstown had all the qualities necessary for getting the maximum services out of his forces, still we believe that he owed not a little of his success to the chief member of his staff. The chief of staff was an incredibly charming and amazingly competent person with the heart of a woman and all the best mental qualities of a man without any of man's defects – the Right Hand Man of 'Uncle Lewis', otherwise known as the 'Only Niece' – Miss Violet Voysey. To this most admirable hostess of Admiralty House we owe a debt of gratitude not only for keeping 'Uncle Lewis' always in a good humor, but for her sympathetic kindness to all the Yankee people, for her charming hospitality, her excellent afternoon tea and such cakes as could be made on the war ration of sugar.'

Conclusion

By virtue of its geographical situation and in the light of the technology available at the time, Bentra had played a small but vital part in the network of airship stations and naval bases around the coast of the British Isles, which had contributed so importantly to winning the first anti-submarine war. It was a welcome haven for the courageous and skilful airship crews in their frail craft. No doubt they enjoyed the hospitality that the kindly folk of Whitehead and Islandmagee would have given to strangers in their midst. This time has all but passed from living memory; it is important therefore that the exploits of the airmen of Bentra should be recorded for posterity.

The contribution of the part played by the Irish Naval Command in winning the war at sea was well summarised in *The Times* of 17th June 1919, 'The part that Queenstown played during the Great War is not generally known, by reason of the very necessary veil of secrecy that was drawn over its work. Some day the full story will be told, and it will prove to be thrilling and amazing, for some of the most daring exploits were performed by the craft based on this port.' To which may be added a comment from the Governor of New York, former Assistant Secretary of the Navy and future President of the United States, Franklin D Roosevelt, who made an official visit to Queenstown on 24th July 1918 on board USS *Kimberley*, Commander AW Johnson, 'When history is finally written, the Queenstown Command will stand out as the finest example of the right spirit of co-operation between our two great countries.' He described his visit as, 'the high spot of my round of inspections of American naval activities in European waters … it is not an easy thing to command an international force. Many young officers came to me in the Navy department and pleaded, almost with tears in their eyes, for assignment to new destroyers that were about to go into commission for duty under the Queenstown command.'

In conclusion, with regard to lighter-than-air aviation, a total of 147 Submarine-Scout type airships were constructed, 29 with the BE2c fuselage, 26 Maurice Farman types, 10 with an Armstrong-Whitworth car, 6 SS-Pushers and 76 SS Zeros. No other contemporary aircraft could have performed the jobs the SS and SSZ airships undertook. None could match the airships' endurance or slow speed capability. Their deterrent value was immense – during the entire war there was only one instance of a ship being escorted by an airship being sunk. This may be placed in context by considering the fact that of the 12,850,815 tons of merchant shipping lost in the Great War, 11,135,460 tons were sunk by U-boats – 88% of the total. Over 5000 ships were sunk with the loss of 15,000 lives. 375 U-boats put to sea, 202 boats and over 5000 sailors did not return (515 officers and 4894 men); 178 were sunk by the Allies and a further 24 were lost due to mechanical failure or accident.

Between April 1917 and October 1918 U-boats sank 1757 ships. 1500 of these were sailing independently, an attrition rate of 5.93%. In contrast only 257 ships sailing in convoy were sunk out of 83,958 convoyed sailings, an attrition rate of 0.3%. In the spring of 1917 the U-boats had been destroying one enemy or merchant ship for every two days spent on patrol. By the early summer of 1918 the average had dropped to one ship for every 14 days of patrol. If it had not been introduced, albeit belatedly, it is highly probable that Britain would have had to sue for peace. A quote from US Admiral Sims is telling, 'Could Germany have kept fifty submarines constantly at work on the great shipping routes in the winter and spring of 1917, nothing could have prevented her from winning the war.'

A report entitled *Summary of Activities of U.S. Naval Forces Operating in European Waters* in 1919, prepared in London shows that the destroyers and other units operating from Irish bases did the jobs assigned to them with considerable efficiency. The Queenstown naval command was the second largest USN base in the war; only Brest, the centre of troop convoy activity, was greater. There were almost 7000 USN personnel in Ireland at the end of the war, and 59 vessels of all types. In the 20 months, the forces in Queenstown supplied 91% of the escorts for 360 convoys. The success of the convoy system can be judged by the sinkings. The monthly shipping tonnage losses tell their own story:

Month	1914	1915	1916	1917	1918
January		47,981	81,259	368,521	306,658
February		59,921	117,547	540,006	318,957
March		80,775	167,097	593,841	342,597
April		55,725	191,667	881,027	278,719
May		120,058	129,175	596,629	295,520
June		131,428	108,851	687,507	255,587
July		109,640	118,215	557,988	260,967
August	62,767	185,866	162,744	511,730	283,815
September	98,378	151,884	230,460	351,748	187,881
October	87,917	88,534	353,660	458,558	118,559
November	19,413	153,043	311,508	289,212	17,682
December	44,197	123,141	355,139	399,212	
Total	**312,672**	**1,307,996**	**2,327,326**	**6,235,878**	**2,666,942**

Grand Total 12,850,815 gross tons

The airships at Luce Bay and Bentra played their part, flying just 33 hours in 1915, but increasing to 405 in 1916, 1757 in 1917 and 4432 in 1918. The comparable totals for all RNAS airships were 339 hours in 1915, 7078 in 1916, 22,389 in 1917 and 53,554 in 1918. During the final 15 months of the war SS type airships carried out over 10,000 patrols, flying nearly one and a half million miles in more than 60,000 hours. Forty-nine U-boats were sighted and 27 of these were attacked from the air or by ships. The submarines were kept below the waves, where they used up valuable battery power and were restricted to a speed of only 8 or 9 knots. A brief log entry from a captured U-boat speaks volumes, 'Sighted airship – submerged.'

Appendix A
RN and USN vessels based in Ireland 1914–18

1. Larne 1914–18

Depot Ship Cruisers

HMS *Hermione*, HMS *Thetis*

North Channel Patrol and Examination Ship

HMS *Tara*

Destroyers

HMS *Avon*, HMS *Dee*, HMS *Dove*, HMS *Express*, HMS *Garry*, HMS *Osprey*, HMS *Thorn*, HMS *Wolf*

Armed Yachts

HMY *Albion III*, HMY *Clementina*, HMY *Jeanette*, HMY *Marynthea*, HMY *Medusa*, HMY *Monsoon*, HMY *Narcissus*, HMY *Oriana*, HMY *Sapphire*, HMY *Valiant*, HMY *Zara*

Armed Trawlers

Alsatian Minor, Angerton, Berkshire, Bittern II, Cerealia, Ceresia, Davara, Diver, Dragon, Earl Lennox, Fishtoft, Glenroy, Goshawk II, Hungarian, Liberia, Margaret Duncan, Nellie Braddock, Neptunian, Prince Victor, Revello, Riano, Rose II, Roxano, Strathmartin, Vera, Vesper, Waltham, War Lord

Armed Drifters

Albatross, Alert, Alfred, Anchor of Hope, Annie Cumine, Annie Smith, Arthur H Johnston, Au Retour, Auch Meddon, Auld Lang Sang, Baden Powell, Bartonia, Belle O'Moray, Betsy Slater, Boy Alan, Boy George, Boy Scout, Breadwinner, Brockhead, Charles Hay, Chatterino, Childrens Trust, Christobel, Christina Mayes, Citron, Cluney Hill, Comely Bank, Convallaria, Correopsis, Coulard Hill, County of Nairn, Craig Min, Culpea, Daffodil, Daisy II, Dashing Spray, David B Summers, Dick Whittington, Do Well, Duthie, Ebenezer P, Effort, Elgar, Elinka, Eminent, Enterprise, Erin III, Explorator, Fanny Mair, Fear Not II, Fern, Ferndale, Fleetwing, Flower, Fortitude, Forward, George Hay, George Walker, Girl Marjory, Girl Winifred P, Gladys & Rose, Gleam of Hope, Glen Urquhart, GMH, GMV, Golden Chance, Golden Dawn, Golden News, Golden Strand, Golden West, Gowan, Grateful, Grey Dawn, Heather Bell, HFE, Holly II, Homeland, Hopeful, Isco, JAC, Jeannie Robertson, Kipper, Laigh O'Moray, Lassie II, Lea Rig, Letterfourie, Liberty II, Light, Lily Oak, Livelihood, Lizzie Brown, Lizzie Hutt, Lloyd George, Loch na Boo, Lonicera, Look Sharp, Lucania II, Maggie Bruce, Maggies, Mariner IV, Maritana, Maryland, Mascot, Masterpiece, Melinka, Midas, Mill Burn, Mormond Hill, Nellie Reid, Nil Desperandum, Northern Scot,

Ocean Harvest, Ocean Pride, Ocean Retriever, Ocean Searcher, Olive, Olive Leaf, Orion, Pilot Me, Pitgaveney, Pitullie, Pride of Buchan, Pride of Moray, Progress, Prospective, Protect Me, Quarry Knowe, Rannas, Redrift, Replenish, Robin, St Combs, Scotsman, Sheila, Silvery Dawn, Snowdrop, Speculation, Speedwell V, Sprig O'Heather, Spring Flower, Suffolk County, Summerton, Swallow, Swift Wing, Tea Rose, Ten, The Brae, The Majesty, The Princess, Thermopylae, Three, Three Kings, Thyrsus, Transit, Triumph, Trophy, Troup Head, Trustful, Two, Tynet, Uberous, Uberty, Valorous, Vesper Star, Victory, Vigorous, Violet Flower, Welcome Home, Welland, W Elliott, Willow Bank, Winner, Young Fred

Patrol Tugs
Alexandra, Blackcock, Harrington, Herculaneum, Hornby, Wallasey

2. Buncrana 1914–18

Armed Trawlers
Alpha, Athelstan, Bedouin, Cerealia, Confederate, Corientes, Denis Casey, Eric Stroud, Erillus, Ferriby, Filey, Grimsby, Helcia, Imperial Queen, Joseph & Sarah Miles, Kaphreda, Liberia, Libyan, Lobelia II, Loch Doon, Lord Knollys, Lord Lister, Lord Tummel, Nathaniel Cole, Nogi, Oakwold, Onetos, Raymont, Revello, Riano, Ribble II, St Johns, Sapphire, Saxon, Scot, Sethon, Strathlui, Thomas Collard, Vale of Lennox, Victoria Regina, War Duke, War Lord, William Biggs, Yucca

Armed Drifters
Alfred, Baden Powell, Bartonia, David B Summers, Emblem, Fancy, Forward II, Lizzie Birrell, Ocean Hope, Peace, Prosit, Snowdrop, Spring Flower, Strathmore, Verdant, Wheatstalk

Destroyers 1917
HMS *Medina*, HMS *Orestes*, HMS *Orford*, HMS *Plucky*

Sloops from 1917
HMS *Buttercup*, HMS *Gladiolus*, HMS *Poppy*, HMS *Rosemary*, HMS *Primrose*, HMS *Laburnum*

Q-ship
Chagford

3. Kingstown

Irish Sea Hunting Flotilla (Kingstown and Holyhead) 1917-18,
HMS *Earnest*, HMS *Griffon*, HMS *Kestrel*, HMS *Lively*, HMS *Mallard*, HMS *Seal*, HMS *Orwell*, HMS *Sprightly*, HMS *Stag*, HMS *Zephyr*

Armed Yachts
HMY *Boadicea II*, HMY *Branwen*, HMY *Greta*, HMY *Helga*, HMY *Ilex*

Armed Trawlers
Dale, Deliverer, Henry Ford, Onyx II, Persian Empire, Presidency, Roman Empire, Sealark II

4. Berehaven, Killybegs, Queenstown, Buncrana

RN Submarines based in Ireland 1917–18 (The first to be stationed at Berehaven in March 1917 are shown in bold type)
C-7, D-1, **D-3**, D-4, D-6, **D-7**, **D-8**, E-23, E-27, **E-32**, E-35, E-43, E-47, E-48, **E-54**, E-56, G-6, **H-5**, H-8, L-1, L-2, L-7, R-7, R-8, R-11, R-12

Submarine Depot Ships
HMS *Ambrose*, HMS *Platypus*, HMS *Vulcan*

5. Killybegs

Armed Yacht
Ellida

Armed Trawlers
Angle, Aquamarine, Craig Island, Eric Stroud, Grackle, Kennymore

Armed Drifters
Thrush, Victoria

6. Berehaven

Armed Yachts
Aster, Greta

Armed Trawlers
Bempton, Brock, Carieda, Drake II, Flying Wing, Ina William, Lord Durham, Lucida, Luneda, Morococala, Reindeer II, Reporto, Vindelicia

Armed Drifters
Daisy VI, Golden Effort, JECM, Lyre Bird, Silvery Dawn

Hospital Ship
Queen Alexandra

Rescue Tugs
Cartmel, Cynic, Dreadful, Flying Foam, Flying Spray

7. Queenstown

11th Cruiser Squadron West of Ireland Coast Patrol 1914–15
HMS *Doris*, HMS *Drake*, HMS *Isis*, HMS *Juno*, HMS *Minerva*, HMS *Sutlej*, HMS *Venus*

Cruisers 1915–18
HMS *Adventure*, HMS *Active*

Destroyers 1915–17
HMS *Magic*, HMS *Marne*, HMS *Mary Rose*, HMS *Narwhal*, HMS *Parthian*, HMS *Peyton*, HMS *Rigorous*, HMS *Sarpedon*, HMS *Taurus*

Sloops 1915–18
HMS *Alyssum*, HMS *Arbutus*, HMS *Aubretia*, HMS *Begonia*, HMS *Bluebell*, HMS *Buttercup*, HMS *Camelia*, HMS *Candytuft*, HMS *Crocus*, HMS *Daffodil*, HMS *Delphinium*, HMS *Flying Fox*, HMS *Genista*, HMS *Heather*, HMS *Iris*, HMS *Jessamine*, HMS *Laburnum*, HMS *Lavender*, HMS *Mignonette*, HMS *Myosotis*, HMS *Poppy*, HMS *Primrose*, HMS *Rhododendron*, HMS *Rosemary*, HMS *Salvia*, HMS *Sir Bevis*, HMS *Snowdrop*, HMS *Sunflower*, HMS *Tamarisk*, HMS *Tulip*, HMS *Viola*, HMS *Zinnia*

Armed Yachts
Adventuress, Aster II, Beryl, Calanthe, Greta, Pioneer II, Scadaun

Patrol and Examination Vessels
HMS *Safeguard*, HMS *Argon*, HMS *Julia*

Seaplane Carrier
HMS *Empress*

Q-Ships 1915–18
Showing Q designation, original name and [in brackets] subsequent names. Haulbowline Dockyard, in Cork Harbour was responsible for the conversion of many merchant vessels to armed decoy ships.
Q 1 *Perugia*, [*Moeraki*]; Q 2 *Intaba*, [*Waitono*], [*Waitopo*]; Q 3 *Barranca*, [*Echunga*]; Q 4 *Carrigan Head*, [*Carrington Head*]; Q 5 *Loderer*, [*Farnborough*]; Q 6 *Zylpha*; Q 7 *Penhurst*, [*Manford*]; Q 8 *Vala*; Q 10 *Begonia*, [*Dolcis*], [*Jessop*]; Q 11, *Tamarisk*; Q 12 *Tulip*; Q 13 *Aubretia*, [*Kai*], [*Winton*], [*Zebal*]; Q 14 *Viola*, [*Damaris*], [*Cranford*]; Q 15 *Salvia*; Q 16 *Heather*, [*Bywater*], [*Lizette*], [*Seetrus*]; Q 25 *Lady Patricia*, [*Anchusa*], [*Paxton*], [*Tosca*]; Q 34 *Acton*, [*Harelda*], [*Woffington*], [*Gandy*]; No 'Q' number, *Vittoria*, [*Pargust*]; No 'Q' number, *Jurassic*, [*Westphalia*], [*Cullist*], [*Hayling*], [*Prim*]; No 'Q' number, *Stonecrop*, [*Ravenstone*], [*Glenfoyle*], [*Dunlevon*]; No 'Q' number, *Arvonian* [Santee] [only US Navy Q-Ship of WW1]; No 'Q' number, *Baron Rose*

(8th) Sweeping Flotilla 1917–18
HMS *Cattistock*, HMS *Cottesmore*, HMS *Epsom*, HMS *Eridge*, HMS *Haldon*, HMS *Hurst*, HMS *Meynell*, HMS *Sandown*, HMS *Southdown*

Armed Trawlers
Ben Gairn, Bluebell, Bradford, Brock, Carieda, Clifton, Concord III, Congo, Ebro, Flying Wing, Freesia, George Milburn, Grenadier, Guillemot, Helios, Heron, Indian Empire, James Johnson, James Seckar, Loch Eye, Lord Durham, Lord Hardinge, Lord Heneage, Maximus, Morococala, Norbreck, Ocean Scout, Oropesa, Reliance II, Reporto, Ristango, Rodney, Sarba, Setter II, Staunton, Thuringia, Venosta, Verbena, Vindelicia

Armed Drifters
Connage, Daisy VI, Expectation, Girl Marjorie, Girl Mary, Golden Effort, Good Hope III, Integrity, JECM, Lily Oak, Monarch II, Morning Star IV, Ocean Angler, Rival II, Speedwell V, Sublime II, Sunshine

Rescue Tugs
Flying Falcon, Flying Foam, Flying Fox, Flying Sportsman, Fylde, Milewater, Paladin II, Stormcock, Warrior

Motor Launch Patrol
ML131, ML161, ML163, ML169, ML171, ML173, ML183, ML185, ML189, ML197, ML251, ML259, ML374, ML375, ML376, ML377, ML378, ML380, ML381, ML406, ML408, ML412, ML520, ML515, ML524

Hydrophone Flotilla
ML132, ML167, ML181, ML187, ML320, ML325, ML410, ML487

Motorboats
Motorboats were stationed at Baltimore, Dunmore, Kenmare, Kinsale and Castletownbere. *Aptera, California II, Emerald, Mary Rose II, Sakkara, Seagull, Shearwater, Trident*, No 40, No 47, No 85, No 102, No 123, No 124.

8. Londonderry, Buncrana

North Coast of Ireland: 2nd Destroyer Flotilla 1917–18
6 G Class, 14 M Class at Londonderry – HMS *Mandate*, HMS *Marne*, HMS *Martial*, HMS *Michael*, HMS *Milbrook*, HMS *Minos*, HMS *Moresby*, HMS *Nicator*, HMS *Ossory*, HMS *Pelican*, HMS *Pigeon*, HMS *Magic*, HMS *Manners*, HMS *Medway*, HMS *Mindful*, HMS *Mons*, HMS *Mounsey*, HMS *Mystic*, HMS *Musketeer*, HMS *Beagle*, HMS *Foxhound*, HMS *Basilisk*, HMS *Grampus*, HMS *Renard*, HMS *Rattlesnake*; attached – HMS *Express*, HMS *Osprey*, HMS *Thorn*, HMS *Wolf*

2nd Sloop Flotilla, 8 sloops at Londonderry – HMS *Ard Patrick*, HMS *Convolvulus*, HMS *Eglantine*, HMS *Marjoram*, HMS *Saxifrage*, HMS *Silene*, HMS *Silvio*, HMS *Spearmint*

9. Rosslare

Hunting Flotilla – Armed Drifters

Expectation, Guide Me II, Sparkling Star, Sublime II

10. Blacksod Bay

Armed Yachts

HMY *Ellida*, HMY *Sanda*

Armed Trawlers

Angle, Craig Island, Grackle, Kennymore, Norbreck

Armed Drifters

Thrush, Victoria

11. Galway

Armed Yachts

HMY *Zarefah*

Armed Trawlers

Alaska, Ben Earn, Craik, Dahlia II, Guillemot, John Chivers, Lord Heneage, Magneta, Ocean Scout, Saxon II, Setter II, Thomas Goble

Armed Drifters

Anchor of Hope, Annie Cummins, Auchmedden, Brighton II, Comely, Expectant, Exuberant, Faithful Friend, Lizzie Hutt, Mary Reid, Melinka, Our Friend, Rosehearty, The Brae, True Friend, Twenty Eight, Welwyn, Wests

Motor Launches

ML374, ML375, ML376, ML380, ML406, ML408, ML 412, ML515, ML524

In 1918 two MLs each were detached to Foynes in Co Limerick and Carrigaholt in Co Clare to watch over the Shannon in co-operation with military cycle patrols. Two more MLs were at both Fenit in Tralee Bay and Westport in Co Mayo, with a further ML in Galway Bay.

US Navy in Ireland 1917–18

12. Berehaven – Battleships

USS *Oklahoma*, USS *Nevada*, USS *Utah*

Berehaven – Submarines

AL-1, AL-2, AL-3, AL-4, AL-9, AL-10, AL-11

13. Queenstown – Destroyers (the first six to arrive are shown in bold type)

USS *Allen*, USS *Ammen*, USS *Aylwin*, USS *Balch*, USS *Beale*, USS *Benham*, USS *Burrows*, USS *Caldwell*, USS *Cassin*, **USS *Conyngham***, USS *Cummings*, USS *Cushing*, **USS *Davis***, USS *Downes*, USS *Drayton*, USS *Duncan*, USS *Ericsson*, USS *Fanning*, USS *Jacob Jones*, USS *Jarvis*, USS *Jenkins*, USS *Kimberley*, USS *McCall*, **USS *McDougal***, USS *Manley*, USS *Nicholson*, USS *O'Brien*, USS *Parker*, USS *Patterson*, USS *Paulding*, USS *Perkins*, **USS *Porter***, USS *Rowan*, USS *Sampson*, USS *Shaw*, USS *Sterett*, USS *Stevens*, USS *Stockton*, USS *Terry*, USS *Trippe*, USS *Tucker*, **USS *Wadsworth***, **USS *Wainwright***, USS *Walke*, USS *Warrington*, USS *Wilkes*, USS *Winslow*

Queenstown – Submarine Chasers

USSC 1, USSC 44, USSC 45, USSC 46, USSC 47, USSC 48, USSC 91, USSC 110, USSC 164, USSC 178, USSC 181, USSC 182, USSC 206, USSC 207, USSC 208, USSC 220, USSC 221, USSC 222, USSC 254, USSC 271, USSC 272, USSC 323, USSC 325, USSC 329, USSC 342, USSC 343, USSC 344, USSC 345, USSC 346, USSC 356

Queenstown – Destroyer and Submarine Repair Ships/Tenders

USS *Melville*, USS *Dixie*, USS *Vestal*, USS *Bushnell*

Naval Tugs

Genesee, Ontario, Sonoma

Armed Yacht

USS *Corsair*

Q Ship

USS *Santee*

Appendix B

Numbers of Available Vessel Types in Irish Auxiliary Patrol Areas

Area XVI KINGSTOWN – HMS BOADICEA II

1st January each year	1915	1916	1917	1918	1919
Yachts	-	3	2	3	3
Trawlers	-	12	12	19	19
Drifters	-	12	12	14	20
Motor Launches	-	-	12	12	12
Motor boats	-	1	-	-	-
Boom defence	-	-	-	1	1
Minesweepers	-	-	-	-	1
Totals	-	28	38	49	56

Area XVII LARNE LOUGH – HMS THETIS

1st January each year	1915	1916	1917	1918	1919
Yachts	2	1	2	3	3
Trawlers	28	18	17	26	33
Drifters	3	107	95	69	45
Motor Launches	-	-	18	12	5
Motor boats	4	3	-	-	-
Boom defence	-	-	-	1	1
Minesweepers	-	- .	-	4	5
Totals	37	129	132	115	92

Area XVIII LOUGH SWILLY – HMS COLLEEN

1st January each year	1915	1916	1917	1918	1919
Yachts	1	-	-	-	-
Trawlers	-	13	20	28	26
Drifters	4	2	3	15	17
Boom defence	-	7	4	4	4
Minesweepers	7	-	-	-	-
Totals	12	22	27	47	47

Area XIX KILLYBEGS – HMS COLLEEN

(5.18 – Became Area XIXA with changed boundaries)

1st January each year	1915	1916	1917	1918	1919
Trawlers	-	7	7	-	3
Totals	-	7	7	-	3

Area XX GALWAY BAY – HMS COLLEEN

(5.18 – Reduced in size and renumbered as new Area XIX)

1st January each year	1915	1916	1917	1918	1919
Yachts	1	-	-		
Trawlers	-	7	7	6	6
Motor Launches	-	-	-	8	9
Totals	1	7	7	14	15

Area XX BEREHAVEN – HMS COLLEEN

(c5.18 – Became new Area XX)

1st January each year	1915	1916	1917	1918	1919
Yachts	1	-	-	-	-
Trawlers	2	-	-	8	13
Drifters	2	-	-	4	5
Motor Launches	-	-	-	6	6
Boom defence	-	-	-	1	4
Totals	5	-	-	19	28

Area XXI QUEENSTOWN – HMS COLLEEN

1st January each year	1915	1916	1917	1918	1919
Yachts	1	4	4	1	-
Trawlers	10	24	24	15	11
Drifters	2	9	10	17	17
Motor Launches	-	-	24	16	15
Motor boats	4	5	-	-	-
Boom defence	-	3	5	5	5
Minesweepers	-	-	-	7	4
Totals	17	45	67	61	52

Bibliography

Official Publications and Files

ADM1/8505/263
ADM1/8508/285
ADM1/8517/70
ADM1/8549/13
ADM137/440
ADM137/441
ADM137/521
ADM137/533
ADM137/534
ADM137/536
ADM137/537
ADM137/540
ADM137/562
ADM137/573
ADM137/592
ADM137/593
ADM137/622
ADM137/623
ADM137/624
ADM137/623
ADM137/650
ADM137/651
ADM137/672
ADM137/673
ADM137/808
ADM137/946
ADM137/947
ADM137/974
ADM137/976
ADM137/1057
ADM137/1128
ADM336/11
ADM340
Air1/270/15/226/111
Air1/419/15/245/1
Air1/420/15/246/1
Air1/420/15/246/2
Air1/434/15/269/1
Air1/434/15/269/2
Air1/442/15/303/14
Air1/443/15/303/4
Air1/481/15/312/249
Air1/485/15/312/269
Air1/485/15/312/270
Air1/486/15/312/271
Air1/489/15/312/280
Air1/489/15/312/281
Air1/489/15/312/282
Air1/627/17/113
Air1/636/17/122/131
Air1/640/17/122/200
Air1/641/17/122/217
Air1/643/17/122/217
Air1/659/17/122/614
Air1/669/17/122/782
Air1/671/17/134/4
Air1/1714/204/123/128
Air2/127
Air3/19B Airship Log Book SS17
Air3/ Airship Log Book SS20
Air3/27, 28, 29, 30, 31 Airship Log Book SS23
DSIR23/9391
Commanding Officer Naval Air Stations Ireland, report on Lough Foyle NAS to Assistant Secretary of the Navy dated 27th January 1919
Flying Officers of the USN, Naval Aviation War Book Committee, Washington DC
Handbook of SS Type Airships. Compiled by the Instructional Staff at the Airship Depot, Wormwood Scrubs, 1917

Letters of Phillip J Gallagher 1918–19
McCrary, Commander FR, History of the US Naval Air Stations, Ireland
Navy List Supplement September 1914–July 1919
RNAS Pilot's Flying Log Book – Lieutenant AH Crump RAF
RNAS Pilot's Flying Log Book – Flight Sub-Lieutenant BW Hemsley RN
Signal Log Book Lough Foyle NAS 'In'
Signal Log Book Lough Foylc NAS 'Out'
Telephone Log Book Lough Foyle NAS 'Odd Dates'
Telephone Log Book Lough Foyle NAS 'Even Dates'
War Diary and History of Lough Foyle NAS
Zabriskie, Ensign GG, Summary of Operations USN Air Station Lough Foyle 1919

Books

Abbott, Patrick, *The British Airship at War 1914-1918* (Lavenham 1989)
Abbott, Patrick, *Airships* (Princes Risborough 1991)
Abbott, Patrick and Walmsley, Nick, *British Airships in Pictures* (Trowbridge 1998)
Barry, Dr JM, *Old Glory at Queenstown* (Cork 1999)
Bayly, Admiral Sir Lewis, *Pull Together!* (London 1939)
Bennett, Geoffrey, *Naval Battles of the First World War* (Barnsley 2005)
Bilbé, Tina, *Kingsnorth Airship Station* (Stroud 2013)
Blazich, Frank A, *United States Navy and World War I: 1914–1922* (Washington DC 2016)
Connon, Peter, *An Aeronautical History of the Cumbria, Dumfries and Galloway Region Part 2 1915–1930* (Penrith 1984)
Campbell, Rear Admiral Gordon, VC, DSO, *My Mystery Ships* (London 1928)
Chatterton, E Keble, *Q-Ships and Their Story* (London 1922)
Chatterton, E Keble, *The Auxiliary Patrol* (London 1923)
Chatterton, E Keble, *Danger Zone – The Story of the Queenstown Command* (London 1934)
Churchill, Winston S, *The Great War* (London 1933)
Corbett, Sir Julian S, *History of the Great War, Naval Operations, Volume II* (London 1921)
Compton-Hall, Richard, *Submarines and the War at Sea 1914–18* (London 1991)
Cork County Council, *A Commemoration of the US Navy in Cork* (Cork 2017)
Crawford, WH (editor), *Industries of the North* (Belfast 1986)
Elmhirst, Sir Thomas, *Recollections of Air Marshal Sir Thomas Elmhirst*, (Privately published, 1991)
Fife, Malcolm, *British Airship Bases of the 20th Century* (Oxford 2013)
Fitzgerald, Pat, *Down Paths of Gold* (Cork 1992)
Gaul, Liam, *Wings Over Wexford* (Dublin 2017)
Gibson, RH and Prendergast, Maurice, *The German Submarine War 1914–1918* (Uckfield 2003)
Gray, Edwin, *A Damned Un-English Weapon* (London 1971)
Gray, Edwin, *The Killing Time* (London 1975)

Gwatkin-Williams, Captain RS, *Prisoners of the red desert; being a full and true history of the men of the 'Tara'* (New York, 1923)
Hayes, Karl E, *A History of the Royal Air Force and United States Naval Air Service in Ireland 1913-1923* (Killiney 1988)
Henry, William, *Galway and The Great War* (Cork 2006)
Higham R, *The British Rigid Airship 1908-1931* (London 1961)
Irish Air Letter, *US Navy Aircraft in Ireland* (Dublin 1996)
Le Fleming, HM, *Warships of World War 1* (London 1959)
Maginnis, Matt *Mourne Men and the U-boats 1914–1918* (Bangor 2013)
Massie, Robert K, *Dreadnought* (London 2004)
Moss, Michael and Hume, John R, *Shipbuilders to the World* (Belfast 1986)
Mowthorpe, Ces, *Battlebags* (Stroud 1995)
Nolan, Liam and Nolan, John E, *Secret Victory – Ireland and the War at Sea 1914–1918* (Cork 2009)
O'Donnell, PJ, *Whitehead the town with no streets* (Belfast 1998)
O'Sullivan, Patrick, *The Sinking of the Lusitania* (Cork 2014)
Pitt, Barry, *Zeebrugge* (London 1958)
Price, Gilbert Holland, *The Innocent Erk – RNAS and RAF Memoirs of Clk 2* (unpublished manuscript RAF Museum)
Raleigh W and Jones AH, *The War in the Air* (six volumes Oxford 1922)
Sims, Rear Admiral William Sowden, *The Victory at Sea* (London 1920)
Stokes, Roy, *Death in the Irish Sea: the sinking of the RMS Leinster* (Cork, 1998)
Sturtivant, Ray and Page, Gordon, *Royal Naval Aircraft Serials and Units* 1911–19 (Tonbridge 1992)
Taussig, Commander Joseph K, *The Queenstown Patrol 1917* (Newport, 1996)
Terraine, John, *Business in Great Waters* (London 1989)
Thomas, Lowell, *Raiders of the Deep* (London 1929)
Thompson, Julian, *The Imperial War Museum Book of the War at Sea 1914–1918* (London 2005)
Thompson, Robert, *Five Brave Giant's Causeway Fishermen and An Act of Great Humanity* (Bushmills 2008)
Tillotson, Lieutenant [jg] CB, *US Naval Air Station Wexford Souvenir* (Wexford 1919)
Treadwell, Terry C, *America's First Air War* (Shrewsbury 2000)
Van Wyen, Adrian O, *Naval Aviation in World War 1* (Washington 1969)
Warner, Guy, *Naval Aviation in Inishowen World War 1* (Greencastle 2018)
Warner, Guy with Boyd, Alex, *Army Aviation in Ulster* (Newtownards 2004)
Warner, Guy and Woods, Jack, *Belfast International Airport – Aviation at Aldergrove Since 1918* (Newtownards 2001)
Williams, Captain TB, *Airship Pilot No.28* (London 1974)
Wilson, Ian, *Shipwrecks of the Ulster Coast* (Coleraine 2000)

Articles

Collins, Timothy, *The 'Helga/Muirchu' Her Contribution to Galway Maritime History* (Journal of the Galway Archaeological and Historical Society Vol 54 2002)

Cummins, Patrick J, *US Naval Air Station Wexford 1918* (*Aviation News* 6th–19th February 1987)

Delany, Vice Admiral Walter S, *Bayly's Navy* (Naval Historical Foundation 1980)

Elmhirst, Air Marshal Sir Thomas (*Aerostat* Vol 8 No 1 1977)

Jackson, Robert, *Airships in the war against the U-boats* (*The Rolls-Royce Magazine*, September 1992)

Fanning, Ronan, *Revealed: US naval air base on Wexford Coast* (*Sunday Independent*, 16th April 2006)

Hamill, Norman, *Memories of Ture seaplane base* (*North West Telegraph* 21st July 2004)

Kinsella, Anthony, *The Royal Navy and the Sinn Fein Rebellion* (Military History Society of Ireland, February 2016)

Rossano, Dr Geoffrey, *US Naval Aviation in Ireland* (*Foundation*, Fall 2011)

Belfast Weekly Telegraph, 30th August 1913

Connacht Tribune, 23rd January 1915

Larne Weekly Telegraph, August 1913

Wigtown Free Press by Donnie Nelson, 26th June 1986, 18th September 1986, 25th January 1996

Larne Times, 5th September 1986

Ulster Airmail by Guy Warner, January 2004

Websites

www.1915crewlists.rmg.co.uk (Merchant Navy)
www.airshipsonline.com (Airship Heritage Trust)
www.clydeships.co.uk
www.corkshipwrecks.net
www.dreadnoughtproject.org
www.flightglobal.com
www.geocities.com (U-boat war 1914-18)
www.gwpda.org/naval (German Admiralty Declaration 4.2.1915 and following events)
www.history.navy.mil (US Naval History and Heritage Command)
www.ibiscom.com/sub.htm (U-boat attack 1916)
www.irishwrecksonline.net
www.libraryireland.com
www.naval-history.net
www.ocotilloroad.com, www.ncbi.nlm.nih.gov (Charles Carpendale)
www.rafweb.org (Air of Authority a History of RAF Organisation)
www.uboat.net
www.vlib.us/wwi/resources/archives/texts/uboatu9.html

Conversation and correspondence by letter, e-mail and telephone

Patrick Abbott, Elaine Barton, Michael Bradshaw, Séan Brennan, Frank Brophy, Mrs Nancy Calwell, Dr Meg Carroll, Ernie Cromie, Tim Elmhirst, Peter Garth, Liam Gaul, Vera Girvan, David Hobbs, Anthony Kinsella, Stuart Leslie, Wilson Logan, Mrs Jane Mackie, Mrs Val MacLeod, Matt Maginnis, Joan Morris, Ces Mowthorpe, Donnie Nelson, Billy Noble, PJ O'Donnell, Seán T Rickard, Michael Traynor, Brian Turpin, John Ware, Norman Whitla, Den Burchmore, Giles Camplin (Airship Heritage Trust), Mick Davis (Cross & Cockade), Bill Addison, Royal Cheney, Marc Levitt, Bob Thomas (National Naval Aviation Museum, Pensacola), Mrs Emma Crocker (Imperial War Museum), Nigel Curtin (Dún Laoghaire Library), Peter Devitt (RAF Museum), Stephen Courtney, Victoria Millar (NMRN), John Duggan (Zeppelin Study Group), Stephen Daye (Mid and East Antrim Borough Council), Mark Frost (Dover Museum), Richard Haigh (Rolls-Royce Heritage Trust), Ivor Hamrock (Mayo County Library), Marian Kelso, Gemma Reid, Chrissie Williamson (Larne Museum), Jonathan Roscoe (Naval History & Heritage Command, Washington Navy Yard), Jerry Shore (FAA Museum)

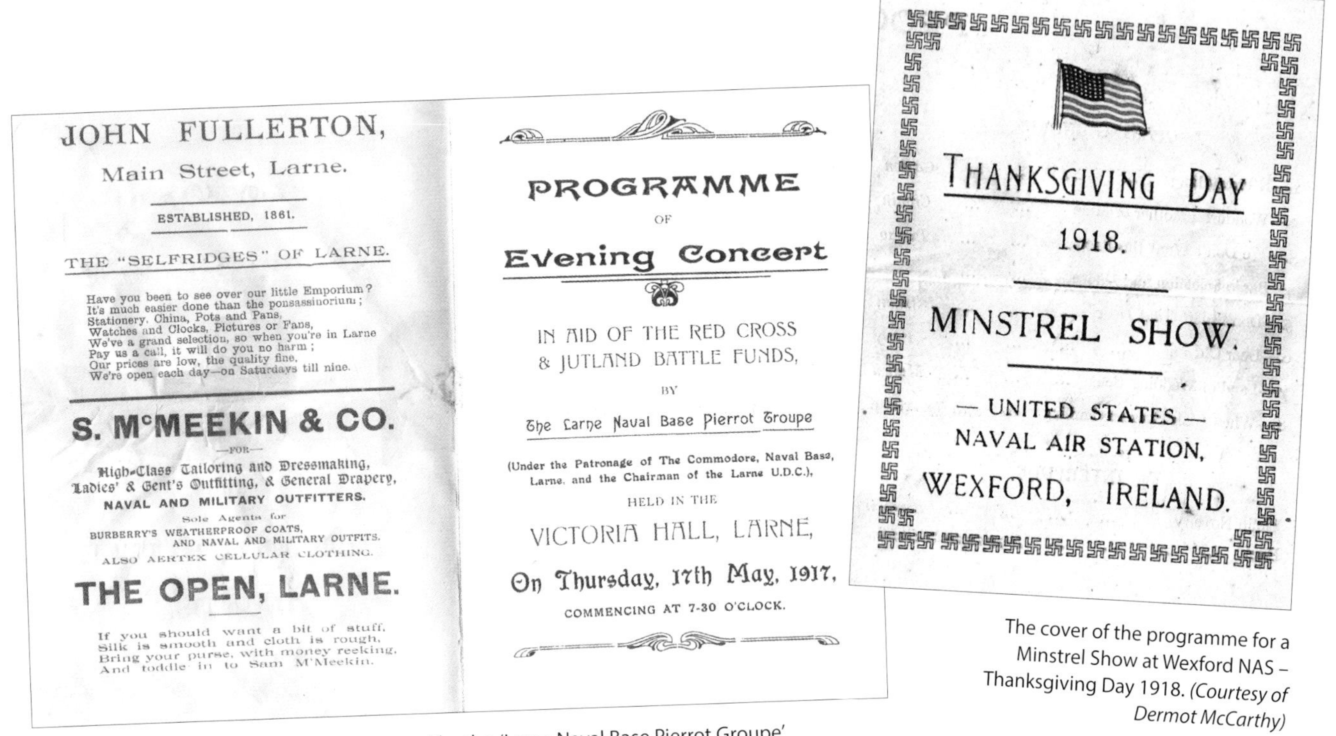

The programme for an evening concert staged by the 'Larne Naval Base Pierrot Groupe' in 1917, in aid of the Red Cross and 'Jutland Battle Funds'. *(Author's Collection)*

The cover of the programme for a Minstrel Show at Wexford NAS – Thanksgiving Day 1918. *(Courtesy of Dermot McCarthy)*

Index

People

Places

Aeroplanes and Airships

Balloons

Ships and other vessels

Miscellaneous